SpringerBriefs in Energy

For further volumes:
http://www.springer.com/series/8903

Mehmet Kanoğlu · Yunus A. Çengel
İbrahim Dinçer

Efficiency Evaluation of Energy Systems

 Springer

Mehmet Kanoğlu
Department of Mechanical Engineering
University of Gaziantep
Gaziantep, Turkey

Yunus A. Çengel
Department of Mechanical Engineering
University of Nevada
Reno, NV, USA

İbrahim Dinçer
Department of Mechanical Engineering
University of Ontario Institute
of Technology
Oshawa, ON, Canada

ISSN 2191-5520   ISSN 2191-5539 (electronic)
ISBN 978-1-4614-2241-9   ISBN 978-1-4614-2242-6 (eBook)
DOI 10.1007/978-1-4614-2242-6
Springer New York Heidelberg Dordrecht London

Library of Congress Control Number: 2012932274

Printed on acid-free paper

Springer is part of Springer Science+Business Media (www.springer.com)

Preface

Efficiency has become a prime topic in energy studies, and has attracted a great deal of attention from all disciplines from mechanical engineering to physics and from electrical engineering to architecture. Efficiency calculations are commonly used to evaluate systems, applications, processes, and services in every sector, ranging from industrial to residential and from utility to commercial. Efficiency evaluation is considered a key component in assessing the performance of energy systems.

The two main principles of thermodynamics are the first and second laws of thermodynamics which lead to energy efficiency and exergy efficiency, respectively. A few decades ago, a thermodynamic analysis involved the evaluation of energy efficiencies only without consideration of second law effects. Growing awareness of limited energy resources and concern for a sustainable economy have made it necessary to perform more refined thermodynamics studies, which required a realistic assessment of the degradation of energy due to irreversibilities. Such a necessity has made exergy efficiency an indispensable tool for performance evaluation.

This book primarily covers energy and exergy efficiencies and their associated discussion, and provides the necessary tools to analyze various systems and applications and to make comparisons. Coverage of the material is extensive, and the amount of information and data presented is sufficient for detailed studies. This book should be of interest to students, researchers, engineers, and practitioners in the area of energy as well as people who are interested in evaluating and improving energy systems and applications. The book should also serve as a valuable reference and source book for anyone who wishes to learn more about efficiency assessment.

The first chapter addresses general aspects of energy, efficiency, environment, and sustainable development as well as linkages between them with some examples. Chapter 2 introduces both the first and second laws of thermodynamics and discusses their role and use in practical applications. Chapter 3 goes further and introduces both energy and exergy efficiencies as valuable tools for performance evaluations and illustrates their use through some examples. It also puts these efficiencies in perspective, and highlights the differences between them. Chapter 4 deals specifically with energy conversion efficiencies and introduces basic formulations for use in common applications. Chapter 5 presents thermody-namic modeling of power plants and their performance assessment using both energy and exergy efficiencies. Finally, analysis and performance assessment is extended in Chap. 6 to refrigeration systems, again through energy and exergy efficiencies. Incorporated throughout this book is a wide range of examples from a diverse area of practical applications.

Gaziantep, Turkey   Mehmet Kanoğlu
Reno, NV, USA   Yunus A. Çengel
Oshawa, ON, Canada   İbrahim Dinçer

Contents

Chapter 1
Efficiency, Environment, and Sustainability

1.1 Introduction

In thermodynamics, the description of an energy conversion system is usually followed by an appropriate efficiency definition of the system. A concentrated study of thermodynamics may be accomplished by the study of various efficiencies and ways to increase them.

For an engineering system, efficiency, in general, can be defined as the ratio of desired output to required input. Although this definition provides a simple general understanding of efficiency, a variety of specific efficiency relations for different engineering systems and operations has been developed. Different efficiency definitions based on the first and second laws of thermodynamics have been the subject of a large number of publications. Various efficiency definitions used for common energy conversion systems are the topic of this book. Many approaches that can be used to define efficiencies are provided and their implications are discussed. This book uses a logical and intuitive approach in defining efficiencies, and it is intended to provide a clear understanding of various efficiencies used in many common energy systems.

1.2 Energy, Exergy, Environment, and Sustainable Development

Using energy-efficient appliances and practicing energy conservation measures help our pocketbooks by reducing our utility bills. They also help the environment by reducing the amount of pollutants emitted to the atmosphere during the combustion of fuel at home or at the power plants where electricity is generated. The combustion of each therm of natural gas produces 6.4 kg of carbon dioxide, which causes global climate change; 4.7 g of nitrogen oxides and 0.54 g of hydrocarbons,

M. Kanoğlu et al., *Efficiency Evaluation of Energy Systems*,
SpringerBriefs in Energy, DOI 10.1007/978-1-4614-2242-6_1,
© Mehmet Kanoğlu, Yunus A. Çengel, İbrahim Dinçer 2012

which cause smog; 2.0 g of carbon monoxide, which is toxic; and 0.030 g of sulfur dioxide, which causes acid rain. Each therm of natural gas saved eliminates the emission of these pollutants while saving $0.60 for the average consumer in the United States. Each kWh of electricity conserved saves 0.4 kg of coal and 1.0 kg of CO_2 and 15 g of SO_2 from a coal power plant [1].

Exergy analysis is useful for improving the efficiency of energy-resource use, for it quantifies the locations, types, and magnitudes of wastes and losses. In general, more meaningful efficiencies are evaluated with exergy analysis rather than energy analysis, because exergy efficiencies are always a measure of how nearly the efficiency of a process approaches the ideal. Therefore, exergy analysis accurately identifies the margin available to design more efficient energy systems by reducing inefficiencies. Many engineers and researchers agree that thermodynamic performance is best evaluated using exergy analysis because it provides more insights and is more useful in efficiency-improvement efforts than energy analysis alone [2].

Measures to increase energy efficiency can reduce environmental impact by reducing energy losses. From an exergy viewpoint, such activities lead to increased exergy efficiency and reduced exergy losses (both waste exergy emissions and internal exergy consumption). A deeper understanding of the relations between exergy and the environment may reveal the underlying fundamental patterns and forces affecting changes in the environment, and help researchers better deal with environmental damage.

The second law of thermodynamics is instrumental in providing insights into environmental impact. The most appropriate link between the second law and environmental impact has been suggested to be exergy, in part because it is a measure of the departure of the state of a system from that of the environment. The magnitude of the exergy of a system depends on the states of both the system and the environment. This departure is zero only when the system is in equilibrium with its environment.

As stated earlier, exergy analysis is based on the combination of the first and second laws of thermodynamics, and can pinpoint the losses of quality, or work potential, in a system. Exergy analysis is consequently linked to sustainability because in increasing the sustainability of energy use, we must be concerned not only with loss of energy, but also loss of energy quality (or exergy). A key advantage of exergy analysis over energy analysis is that the exergy content of a process stream is a better valuation of the stream than the energy content, because the exergy indicates the fraction of energy that is likely useful and thus utilizable. This observation applies equally at the component level, the process level, and the life cycle level. Application of exergy analysis to a component, process, or sector can lead to insights regarding how to improve the sustainability of the activities comprising the system by reducing exergy losses.

Sustainable development requires not just that sustainable energy resources be used, but that the resources be used efficiently. The authors and others feel that exergy methods can be used to evaluate and improve efficiency and thus to improve sustainability. Inasmuch as energy can never be "lost" as it is conserved according to the first law of thermodynamics, whereas exergy can be lost due to internal

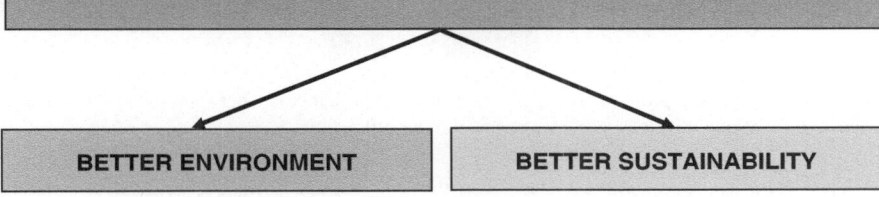

Fig. 1.1 Illustration of how exergy contributes to a better environment and sustainable development

irreversibilities, exergy losses that represent unused potential, particularly from the use of nonrenewable energy forms, should be minimized when striving for sustainable development. Furthermore, Fig. 1.1 clearly summarizes the key advantages of exergy as the potential for a better environment and sustainable development. It is obvious that an understanding of the thermodynamic aspects of sustainable development can help in taking sustainable actions regarding energy. Thermodynamic principles can be used to assess, design, and improve energy and other systems, and to better understand environmental impact and sustainability issues. For the broadest understanding, all thermodynamic principles must be used, not just those pertaining to energy. Thus, many researchers feel that an understanding and appreciation of exergy, as defined earlier, is essential to discussions of sustainable development.

Figure 1.2 illustratively presents the relation among exergy, sustainability, and environmental impact. Here, sustainability is seen to increase and environmental impact to decrease as the process exergy efficiency increases. The two limiting efficiency cases are significant. First, as exergy efficiency approaches 100%, environmental impact approaches zero, because exergy is only converted from one form to another without loss, either through internal consumption or waste emissions. Also sustainability approaches infinity because the process approaches reversibility. Second, as exergy efficiency approaches 0%, sustainability approaches zero because exergy-containing resources are used but nothing is accomplished. Also, environmental impact approaches infinity because, to provide a fixed service, an ever-increasing quantity of resources must be used and a correspondingly increasing amount of exergy-containing wastes are emitted.

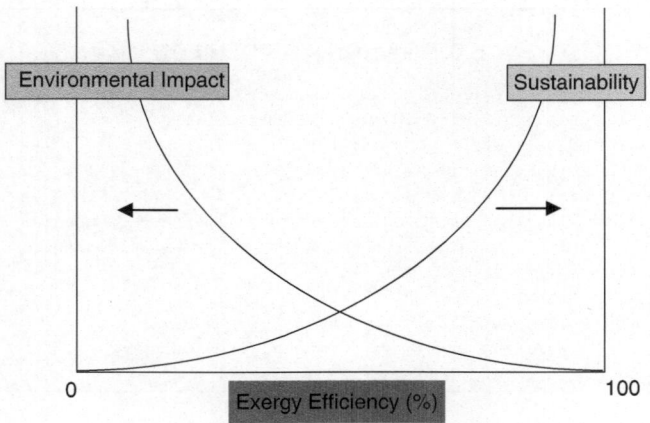

Fig. 1.2 Qualitative representation of the kind of relation that exists between the environmental impact and sustainability of a process, and its exergy efficiency

The relationships between environmental impact and sustainability versus exergy efficiency may be expressed quantitatively by some examples. Further details are available elsewhere [3]. But first we develop the formulation needed for such an analysis. The exergy efficiency of a power plant may be expressed as

$$\varepsilon = \frac{W_{\text{out}}}{X_{\text{in}}} \tag{1.1}$$

where W_{out} is the net work produced and X_{in} is the exergy input, which is equal to the mass of fuel consumed times the specific fuel exergy. The exergy efficiency of a refrigeration cycle is expressible as the actual COP divided by the reversible COP for the same temperature limits:

$$\varepsilon = \frac{\text{COP}_{\text{act}}}{\text{COP}_{\text{rev}}} \tag{1.2}$$

The reversible COP is defined in terms of the temperatures of the low-temperature reservoir T_L and high-temperature reservoir T_H as

$$\text{COP}_{\text{rev}} = \frac{T_L}{T_H - T_L} \tag{1.3}$$

Connelly and Koshland [4] suggest that the efficiency of fossil fuel consumption be characterized by a depletion number defined as

$$D_p = \frac{X_D}{X_{\text{in}}} \tag{1.4}$$

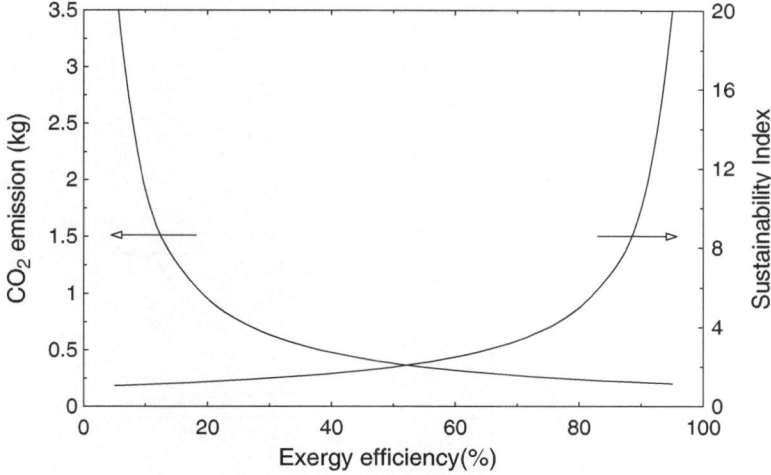

Fig. 1.3 Quantitative illustration of the relation between the carbon dioxide emissions and sustainability index (*SI*) of power generation, and its exergy efficiency. The fuel is methane and the results are for 1 kWh of power output

which represents the relationship between the exergy destruction X_D and the exergy input X_{in} by fuel consumption. The relationship between the depletion factor and the exergy efficiency is

$$\varepsilon = 1 - D_p \tag{1.5}$$

Now, we express the sustainability of the fuel resource by a sustainability index (*SI*) as the inverse of the depletion number:

$$SI = \frac{1}{D_p} \tag{1.6}$$

As a first example, we consider a power plant using natural gas (approximated as methane) as the fuel. We express the environmental impact in terms of the amount of carbon dioxide emissions. A balanced chemical combustion equation of methane shows that for each kilogram of methane burned, 2.75 kg of carbon dioxide (CO_2) is released. The specific chemical exergy of methane is 51,840 kJ/kg [5]. The amount of carbon dioxide emitted and the sustainability index as a function of the exergy efficiency for 1 kWh of power production are plotted in Fig. 1.3. The trends explained in Fig. 1.2 generally apply to the results shown in Fig. 1.3.

As a second example, we consider an air-conditioner used to maintain a space at 25°C (298 K) when the outdoors is at 35°C (308 K). It is assumed that the electricity consumed by this air-conditioner is produced in a coal-fired power plant. For one kilowatt of electricity produced in a coal-fired power plant, 6.38 g of SO_2 and 3.69 g of NO_x are emitted. In this example, we express the environmental impact in terms

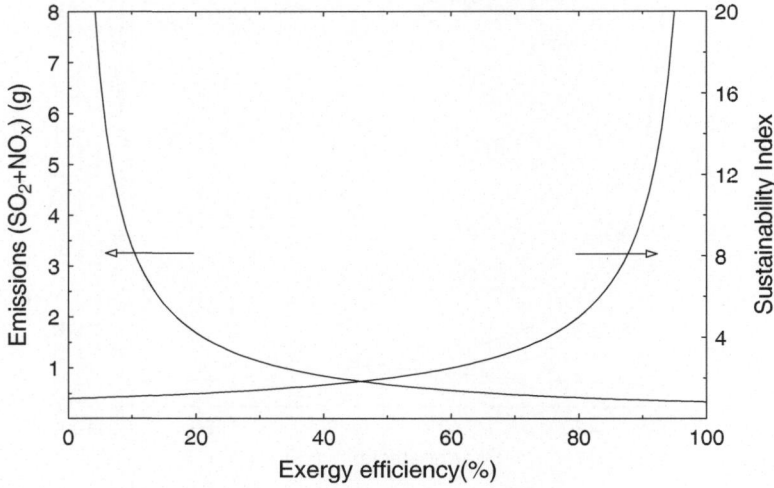

Fig. 1.4 Quantitative illustration of the relation between the SO_2 and NO_x emissions and the sustainability index (*SI*) and its exergy efficiency. The system is an air-conditioner with electricity as the work input. The results are for 1 kWh of cooling load

of the total SO_2 and NO_x emissions. These emissions and the sustainability index as a function of the exergy efficiency for 1 kWh of cooling load from the space are illustrated in Fig. 1.4.

1.3   Efficiency and Energy Management

In the analysis of an energy conversion system, it is important to understand the difference between energy and exergy efficiencies. By considering both of these efficiencies, the quality and quantity of the energy used to achieve a given objective is considered and the degree to which efficient and effective use of energy resources is achieved can be understood. Improving efficiencies of energy systems is an important challenge for meeting energy policy objectives. Reductions in energy use can assist in attaining energy security objectives. Also, efficient energy utilization and the introduction of renewable energy technologies can significantly help solve environmental issues. Increased energy efficiency benefits the environment by avoiding energy use and the corresponding resource consumption and pollution generation. From an economic as well as an environmental perspective, improved energy efficiency has great potential for achieving better sustainability [6].

Accelerated gains in efficiency in energy production and use, particularly in the power generation and utility sectors, can help reduce environmental impact and promote energy security. There is a great technical potential for increased efficiency, however, there exist significant social and economic barriers to its achievement.

Priority should be given to energy policies and strategies that will yield efficiency gains. However, reliance on such policies alone is unlikely to overcome these barriers. For this reason, innovative and bold approaches are required by government, in co-operation with decision makers in the power generation industry, to realize the opportunities for efficiency improvements, and to accelerate the deployment of new and more efficient technologies.

An engineer designing a system is often expected to aim for the highest reasonable technical efficiency at the lowest cost under the prevailing technical, economic, and legal conditions, and with regard to ethical, ecological, and social consequences. Exergy methods can assist in such activities, and offer unique insights into possible improvements. Exergy analysis is a useful tool for addressing the environmental impact of energy resource utilization, and for furthering the goal of more efficient energy resource use, for it enables the locations, types, and true magnitudes of losses to be determined. Also, exergy analysis reveals whether and by how much it is possible to design more efficient energy systems by reducing inefficiencies.

Exergy is also strongly related to sustainability and environmental impact. Sustainability increases and environmental impact decreases as the exergy efficiency of a process increases. As exergy efficiency approaches 100%, the environmental impact associated with process operation approaches zero, because exergy is only converted from one form to another without loss (either through internal consumption or waste emissions). Also the process approaches sustainability because it approaches reversibility. As exergy efficiency approaches 0%, the process deviates as much as possible from sustainability because exergy-containing resources (fuel, ores, steam, etc.) are used but nothing is accomplished. Also, environmental impact increases markedly because, to provide a fixed service, an ever-increasing quantity of resources must be used and a correspondingly increasing amount of exergy-containing wastes is emitted to the surroundings [6, 7].

Energy and exergy efficiencies are considered by many to be useful for the assessment of energy conversion and other systems and for efficiency improvement. However, the use of ambiguous efficiencies that are not clearly defined does not serve this purpose well. A clear, correct, and effective use of energy and exergy efficiencies is crucial in efficiency improvement efforts, which are often a key objective in energy management and policy making.

For governments seeking to improve energy and resource security, by increasing the efficiency with which a society or country uses such resources, exergy provides a critical perspective. It establishes the limits on what can be done and identifies target areas for efficiency improvement (i.e., those areas with high exergy losses). Some work has been done on tracking the exergy flows through regions and economies (e.g., countries, states, provinces). These efforts mainly focus on understanding the true efficiency of energy and resource use in these regions and countries, thereby providing information that is useful to governments and policy makers.

Chapter 2
The First and Second Laws of Thermodynamics

2.1 Introduction

A conventional thermodynamic analysis involves an application of the first law of thermodynamics, also known as energy analysis. Exergy analysis is a thermodynamic analysis technique based on the second law of thermodynamics that provides an alternative and illuminating means of assessing and comparing processes and systems rationally and meaningfully. In particular, exergy analysis yields efficiencies that provide a true measure of how nearly actual performance approaches the ideal, and identifies more clearly than energy analysis the causes and locations of thermodynamic losses and the impact of the built environment on the natural environment. Consequently, exergy analysis can assist in improving and optimizing designs.

Energy and exergy efficiencies are considered by many to be useful for the assessment of energy conversion and other systems and for efficiency improvement. By considering both of these efficiencies, the quality and quantity of the energy used to achieve a given objective is considered and the degree to which efficient and effective use of energy resources is achieved can be understood. Improving efficiencies of energy systems is an important challenge for meeting energy policy objectives. Reductions in energy use can assist in attaining energy security objectives. Also, efficient energy utilization and the introduction of renewable energy technologies can significantly help solve environmental issues. Increased energy efficiency benefits the environment by avoiding energy use and the corresponding resource consumption and pollution generation. From an economic as well as an environmental perspective, improved energy efficiency has great potential [2].

An engineer designing a system is often expected to aim for the highest reasonable technical efficiency at the lowest cost under the prevailing technical, economic, and legal conditions, and with regard to ethical, ecological, and social consequences. Exergy methods can assist in such activities and offer unique insights into possible improvements with special emphasis on environment and

M. Kanoğlu et al., *Efficiency Evaluation of Energy Systems*,
SpringerBriefs in Energy, DOI 10.1007/978-1-4614-2242-6_2,
© Mehmet Kanoğlu, Yunus A. Çengel, İbrahim Dinçer 2012

sustainability. Exergy analysis is a useful tool for addressing the environmental impact of energy resource utilization, and for furthering the goal of more efficient energy-resource use, for it enables the locations, types, and true magnitudes of losses to be determined. Also, exergy analysis reveals whether and by how much it is possible to design more efficient energy systems by reducing inefficiencies.

2.2   Energy Change and Energy Transfer

Energy is the capacity for doing work. The energy of a system consists of internal, kinetic, and potential energies. Internal energy consists of thermal (sensible and latent), chemical, and nuclear energies. Unless there is a chemical or nuclear reaction the internal change of a system is due to thermal energy change. In the absence of electric, magnetic, and surface tension effects, among others, the total energy change of a system is expressed as

$$\Delta E = E_2 - E_1 = \Delta U + \Delta KE + \Delta PE \tag{2.1}$$

where internal, kinetic, and potential energy changes are

$$\Delta U = m(u_2 - u_1) \tag{2.2}$$

$$\Delta KE = \frac{1}{2} m (V_2^2 - V_1^2) \tag{2.3}$$

$$\Delta PE = \frac{1}{2} mg(z_2 - z_1) \tag{2.4}$$

For most cases, the kinetic and potential energies do not change during a process and the energy change is due to the internal energy change:

$$\Delta E = \Delta U = m(u_2 - u_1) \tag{2.5}$$

Energy has the unit of kJ or Btu (1 kJ = 0.94782 Btu). Energy per unit of time is the rate of energy and is expressed as

$$\dot{E} = \frac{E}{\Delta t} \quad (\text{kW or Btu/h}) \tag{2.6}$$

The energy rate unit is kJ/s, which is equivalent to kW or Btu/h (1 kW = 3412.14 Btu/h). Energy per unit mass is called specific energy; it has the unit of kJ/kg or Btu/lbm (1 kJ/kg = 0.430 Btu/lbm).

$$e = \frac{E}{m} \ (\text{kJ/kg or Btu/lbm}) \tag{2.7}$$

Energy can be transferred to or from a system in three forms: mass, heat, and work. They are briefly described below.

2.2.1   Mass Transfer

The mass entering a system carries energy with it and the energy of the system increases. The mass leaving a system decreases the energy content of the system. When a fluid flows into a system at a mass flow rate of $\dot{m}$ (kg/s), the rate of energy entering is equal to mass times energy of a unit mass of a flowing fluid: $\dot{m}(h + V^2/2 + gz)$(kW), where $h = u + Pv$ and Pv is the flow energy (also called flow work) described below.

2.2.2   Heat Transfer

The definitive experiment which showed that heat was a form of energy convertible into other forms was carried out by the Scottish physicist James Joule. Heat is the thermal form of energy and heat transfer takes place when a temperature difference exists within a medium or between different media. Heat always requires a difference in temperature for its transfer. Higher temperature differences provide higher heat transfer rates.

Heat transfer has the same unit as energy. The symbol for heat transfer is Q (kJ). Heat transfer per unit mass is denoted by q (kJ/kg). Heat transfer per unit time is the rate of heat transfer $\dot{Q}$ (kW). If there is no heat transfer involved in a process, it is called an *adiabatic process*.

2.2.3   Work

Work is the energy that is transferred by a difference in pressure or under the effect of a force of any kind and is subdivided into shaft work and flow work. Work is denoted by W. Shaft work is mechanical energy used to drive a mechanism such as a pump, compressor, or turbine. Flow work is the energy transferred into a system by fluid flowing into, or out of, the system. The rate of work transfer per unit time is called *power* $\dot{W}$(kW). Work has the same unit as energy. The direction of heat and work interactions can be expressed by sign conventions or using subscripts such as "in" and "out".

Fig. 2.1 A general closed system with heat and work interactions

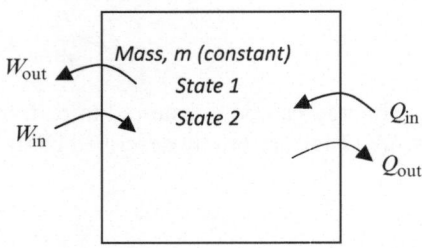

2.3 The First Law of Thermodynamics

So far, we have considered various forms of energy such as heat Q, work W, and total energy E individually, and no attempt has been made to relate them to each other during a process. The first law of thermodynamics, also known as the conservation of energy principle, provides a sound basis for studying the relationships among the various forms of energy and energy interactions.

Based on experimental observations, the first law of thermodynamics states that energy can be neither created nor destroyed during a process; it can only change forms. Therefore, every bit of energy should be accounted for during a process. We all know that a rock at some elevation possesses some potential energy, and part of this potential energy is converted to kinetic energy as the rock falls. Experimental data show that the decrease in potential energy (mgz) exactly equals the increase in kinetic energy when the air resistance is negligible, thus confirming the conservation of energy principle for mechanical energy [1].

The first law of thermodynamics can be expressed for a general system inasmuch as the net change in the total energy of a system during a process is equal to the difference between the total energy entering and the total energy leaving the system:

$$E_{in} - E_{out} = \Delta E_{system} \tag{2.8}$$

In rate form,

$$\dot{E}_{in} - \dot{E}_{out} = \frac{dE}{dt} \tag{2.9}$$

For a closed system undergoing a process between initial and final states (states 1 and 2) with heat and work interactions with the surroundings (Fig. 2.1):

$$E_{in} - E_{out} = \Delta E_{system}$$
$$(Q_{in} + W_{in}) - (Q_{out} + W_{out}) = \Delta U + \Delta KE + \Delta PE \tag{2.10}$$

Fig. 2.2 A general
steady-flow control volume
with mass, heat, and work
interactions

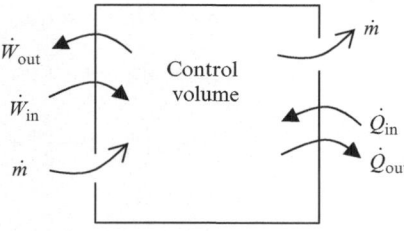

Fig. 2.3 A general
unsteady-flow process
with mass, heat, and work
interactions

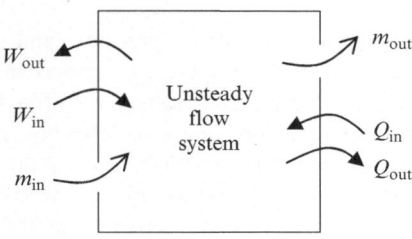

If there is no change in kinetic and potential energies:

$$(Q_{\text{in}} + W_{\text{in}}) - (Q_{\text{out}} + W_{\text{out}}) = \Delta U = m(u_2 - u_1) \tag{2.11}$$

Let us consider a control volume involving a steady-flow process. Mass is entering and leaving the system and there are heat and work interactions with the surroundings (Fig. 2.2). During a steady-flow process, the total mass and energy content of the control volume remains constant, and thus the total energy change of the system is zero. Then the first law of thermodynamics can be expressed through the balance equation as

$$\dot{E}_{\text{in}} - \dot{E}_{\text{out}} = \frac{dE}{dt} = 0$$

$$\dot{E}_{\text{in}} = \dot{E}_{\text{out}}$$

$$\dot{Q}_{\text{in}} + \dot{W}_{\text{in}} + \dot{m}\left(h_{\text{in}} + \frac{V_{\text{in}}^2}{2} + gz_{\text{in}}\right) = \dot{Q}_{\text{out}} + \dot{W}_{\text{out}} + \dot{m}\left(h_{\text{out}} + \frac{V_{\text{out}}^2}{2} + gz_{\text{out}}\right) \tag{2.12}$$

If the changes in kinetic and potential energies are negligible, it results in

$$\dot{Q}_{\text{in}} + \dot{W}_{\text{in}} + \dot{m}h_{\text{in}} = \dot{Q}_{\text{out}} + \dot{W}_{\text{out}} + \dot{m}h_{\text{out}} \tag{2.13}$$

Charging and discharging processes may be modeled as unsteady-flow processes. Consider an unsteady-flow process as shown in Fig. 2.3. Assuming uniform flow conditions, the mass and energy balance relations may be expressed as

$$m_{\text{in}} - m_{\text{out}} = m_2 - m_1 \tag{2.14}$$

$$E_{in} - E_{out} = \Delta E_{system}$$

$$Q_{in} + W_{in} + m_{in}\left(h_{in} + \frac{V_{in}^2}{2} + gz_{in} \right) - Q_{out} - W_{out}$$

$$- m_{out}\left(h_{out} + \frac{V_{out}^2}{2} + gz_{out} \right) = m_2 u_2 - m_1 u_1 \qquad (2.15)$$

2.4   The Second Law of Thermodynamics

Energy is a conserved property, and no process is known to have taken place in violation of the first law of thermodynamics. Therefore, it is reasonable to conclude that a process must satisfy the first law to occur. However, as explained here, satisfying the first law alone does not ensure that the process will actually take place.

It is common experience that a cup of hot coffee left in a cooler room eventually loses heat. This process satisfies the first law of thermodynamics because the amount of energy lost by the coffee is equal to the amount gained by the surrounding air. Now let us consider the reverse process: the hot coffee getting even hotter in a cooler room as a result of heat transfer from the room air. We all know that this process never takes place. Yet, doing so would not violate the first law as long as the amount of energy lost by the air is equal to the amount gained by the coffee.

As another familiar example, consider the heating of a room by the passage of electric current through a resistor. Again, the first law dictates that the amount of electric energy supplied to the resistance wires be equal to the amount of energy transferred to the room air as heat. Now let us attempt to reverse this process. It will come as no surprise that transferring some heat to the wires does not cause an equivalent amount of electric energy to be generated in the wires.

Finally, consider a paddle-wheel mechanism that is operated by the fall of a mass. The paddle wheel rotates as the mass falls and stirs a fluid within an insulated container. As a result, the potential energy of the mass decreases, and the internal energy of the fluid increases in accordance with the conservation of energy principle. However, the reverse process, raising the mass by transferring heat from the fluid to the paddle wheel, does not occur in nature, although doing so would not violate the first law of thermodynamics.

It is clear from these arguments that processes advance in a certain direction and not in the reverse direction. The first law places no restriction on the direction of a process, but satisfying the first law does not ensure that the process can actually occur. This inadequacy of the first law to identify whether a process can take place is remedied by introducing another general principle, the second law of thermodynamics. We show later in this chapter that the reverse processes discussed above violate the second law of thermodynamics. This violation is easily detected with the help of a property called *entropy*. A process cannot occur unless it satisfies both the first and the second laws of thermodynamics.

The use of the second law of thermodynamics is not limited to identifying the direction of processes. The second law also asserts that energy has quality as well as quantity. The first law is concerned with the quantity of energy and the transformations of energy from one form to another with no regard to its quality. Preserving the quality of energy is of major concern to engineers, and the second law provides the necessary means to determine the quality as well as the degree of degradation of energy during a process. As discussed later in this chapter, more of high-temperature energy can be converted to work, and thus it has a higher quality than the same amount of energy at a lower temperature.

The second law of thermodynamics is also used in determining the theoretical limits for the performance of commonly used engineering systems such as heat engines and refrigerators, as well as predicting the degree of completion of chemical reactions. The second law is also closely associated with the concept of perfection. In fact, the second law defines perfection for thermodynamic processes. It can be used to quantify the level of perfection of a process, and point out the direction to eliminate imperfections effectively.

Energy has quality as well as quantity. More of the high-temperature thermal energy can be converted to work. Therefore, it results in "the higher the temperature is, the higher the quality of the energy." Large quantities of solar energy, for example, can be stored in large bodies of water called solar ponds at about 350 K. This stored energy can then be supplied to a heat engine to produce work (electricity). However, the efficiency of solar pond power plants is very low (under 5%) because of the low quality of the energy stored in the source, and the construction and maintenance costs are relatively high. Therefore, they are not competitive even though the energy supply of such plants is free. The temperature (and thus the quality) of the solar energy stored could be raised by utilizing concentrating collectors, but the equipment cost in that case becomes very high.

Work is a more valuable form of energy than heat inasmuch as 100% of work can be converted to heat, but only a fraction of heat can be converted to work. When heat is transferred from a high-temperature body to a lower temperature one, it is degraded because less of it can now be converted to work. For example, if 100 kJ of heat are transferred from a body at 1,000 K to a body at 300 K, at the end we will have 100 kJ of thermal energy stored at 300 K, which has no practical value. But if this conversion were made through a heat engine, up to $1 - 300/1000 = 0.70 = 70\%$ of it could be converted to work, which is a more valuable form of energy. Thus 70 kJ of work potential is wasted as a result of this heat transfer, and energy is degraded [1].

There are numerous forms of second law statements. Two classical statements are as follows.

The Kelvin–Plank statement: It is impossible to construct a device, operating in a
   cycle (e.g., heat engine), that accomplishes only the extraction of heat energy
   from some source and its complete conversion to work. This simply shows the
   impossibility of having a heat engine with a thermal efficiency of 100%.
The Clausius statement: It is impossible to construct a device, operating in a cycle
   (e.g., refrigerator and heat pump), that transfers heat from the low-temperature
   side (cooler) to the high-temperature side (hotter), and producing no other effect.

2.5 Entropy

The second law of thermodynamics often leads to expressions that involve inequalities. An irreversible (i.e., actual) heat engine, for example, is less efficient than a reversible one operating between the same two thermal energy reservoirs. Likewise, an irreversible refrigerator or a heat pump has a lower coefficient of performance (COP) than a reversible one operating between the same temperature limits. Another important inequality that has major consequences in thermodynamics is the *Clausius inequality*. It was first stated by the German physicist R.J.E. Clausius (1822–1888), one of the founders of thermodynamics, and is expressed as

$$\oint \frac{\delta Q}{T} \leq 0 \tag{2.16}$$

That is, the cyclic integral of $\delta Q/T$ is always less than or equal to zero. This inequality is valid for all cycles, reversible or irreversible.

Clausius realized in 1865 that he had discovered a new thermodynamic property, and he chose to name this property *entropy*. It is designated S and is defined as

$$dS = \left(\frac{\delta Q}{T}\right)_{\text{int rev}} \tag{2.17}$$

Entropy is an extensive property of a system and is sometimes referred to as *total entropy*. Entropy per unit mass, designated s, is an intensive property and has the unit kJ/kg · K. The term *entropy* is generally used to refer to both total entropy and entropy per unit mass because the context usually clarifies which one is meant.

The entropy change of a system during a process can be determined by integrating (2.17) between the initial and the final states:

$$\Delta S = S_2 - S_1 = \int_1^2 \left(\frac{\delta Q}{T}\right)_{\text{int rev}} \tag{2.18}$$

Note that entropy is a property, and like all other properties, it has fixed values at fixed states. Therefore, the entropy change between two specified states is the same no matter what path, reversible or irreversible, is followed during a process.

Consider a cycle made up of two processes: process 1-2, which is arbitrary (reversible or irreversible), and process 2-1, which is internally reversible. From the Clausius inequality it becomes

$$\int_1^2 \left(\frac{\delta Q}{T}\right) + \int_2^1 \left(\frac{\delta Q}{T}\right)_{\text{int rev}} \leq 0 \tag{2.19}$$

The second integral in the previous relation is recognized as the entropy change $S_1 - S_2$. Therefore,

$$\int_1^2 \left(\frac{\delta Q}{T}\right) + S_1 - S_2 \leq 0 \tag{2.20}$$

which can be rearranged as

$$S_2 - S_1 \geq \int_1^2 \left(\frac{\delta Q}{T}\right) \tag{2.21}$$

The inequality sign in the preceding relations is a constant reminder that the entropy change of a closed system during an irreversible process is always greater than the entropy transfer. That is, some entropy is generated or created during an irreversible process, and this generation is due entirely to the presence of irreversibilities. The entropy generated during a process is called *entropy generation* and is denoted by Sgen. Noting that the difference between the entropy change of a closed system and the entropy transfer is equal to entropy generation, (2.21) can be rewritten as an equality as

$$\Delta S_{\text{sys}} = S_2 - S_1 = \int_1^2 \left(\frac{\delta Q}{T}\right) + S_{\text{gen}} \tag{2.22}$$

Note that the entropy generation S_{gen} is always a positive quantity or zero. Its value depends on the process, and thus it is not a property of the system. Also, in the absence of any entropy transfer, the entropy change of a system is equal to the entropy generation.

Equation 2.22 has far-reaching implications in thermodynamics. For an isolated system (or simply an adiabatic closed system), the heat transfer is zero, and (2.21) reduces to

$$\Delta S_{\text{isolated}} \geq 0 \tag{2.23}$$

This equation can be expressed as the entropy of an isolated system during a process always increases or, in the limiting case of a reversible process, remains constant. In other words, it never decreases. This is known as the *increase of entropy principle*. Note that in the absence of any heat transfer, entropy change is due to irreversibilities only, and their effect is always to increase entropy.

Entropy is an extensive property, thus the total entropy of a system is equal to the sum of the entropies of the parts of the system. An isolated system may consist of any number of subsystems. A system and its surroundings, for example, constitute an isolated system inasmuch as both can be enclosed by a sufficiently large arbitrary boundary across which there is no heat, work, or mass transfer. Therefore, a system and its surroundings can be viewed as the two subsystems of an isolated system, and the entropy change of this isolated system during a process is the sum of the entropy changes of the system and its surroundings, which is equal to the entropy generation because an isolated system involves no entropy transfer. That is,

$$S_{gen} = \Delta S_{total} = \Delta S_{sys} + \Delta S_{surr} \geq 0 \qquad (2.24)$$

where the equality holds for reversible processes and the inequality for irreversible ones. Note that ΔS_{surr} refers to the change in the entropy of the surroundings as a result of the occurrence of the process under consideration. No actual process is truly reversible, and so we can conclude that some entropy is generated during a process, and therefore the entropy of the universe, which can be considered to be an isolated system, is continuously increasing. The more irreversible a process is, the larger the entropy generated during that process. No entropy is generated during reversible processes.

The increase of the entropy principle does not imply that the entropy of a system cannot decrease. The entropy change of a system can be negative during a process, but entropy generation cannot. The performance of engineering systems is degraded by the presence of irreversibilities, and entropy generation is a measure of the magnitudes of the irreversibilities present during that process. The greater the extent of irreversibilities is, the greater the entropy generation. Therefore, entropy generation can be used as a quantitative measure of irreversibilities associated with a process. It is also used to establish criteria for the performance of engineering devices.

2.5.1   Entropy Balance

The entropy property is a measure of molecular disorder or randomness of a system, and the second law of thermodynamics states that entropy can be created but it cannot be destroyed. Therefore, the entropy change of a system during a process is greater than the entropy transfer by an amount equal to the entropy generated during the process within the system, and the increase of entropy principle for any system is expressed as

$$S_{in} - S_{out} + S_{gen} = \Delta S_{system} \qquad (2.25)$$

This relation is often referred to as the *entropy balance* and is applicable to any system undergoing any process. The entropy balance relation above can be stated as the entropy change of a system during a process is equal to the net entropy transfer through the system boundary and the entropy generated within the system.

Entropy can be transferred by heat and mass. Entropy transfer by heat is expressed as

$$S_{heat} = \frac{Q}{T} \qquad (2.26)$$

Entropy transfer by mass is given by

$$S_{\text{mass}} = ms \qquad (2.27)$$

When two systems are in contact, the entropy transfer from the warmer system is equal to the entropy transfer into the cooler one at the point of contact. That is, no entropy can be created or destroyed at the boundary because the boundary has no thickness and occupies no volume. Note that work is entropy-free, and no entropy is transferred by work. The entropy balance in (2.25) can be expressed in the rate form as

$$\dot{S}_{\text{in}} - \dot{S}_{\text{out}} + \dot{S}_{\text{gen}} = dS_{\text{system}}/dt \qquad (2.28)$$

Let us reconsider the closed system in Fig. 2.1. The entropy balance on this closed system may be written as

$$\left(\frac{Q}{T}\right)_{\text{in}} - \left(\frac{Q}{T}\right)_{\text{out}} + S_{\text{gen}} = m(s_2 - s_1) \qquad (2.29)$$

Now, consider the control volume in Fig. 2.2 with a steady-flow process. The entropy balance on this control volume may be expressed as

$$\left(\frac{\dot{Q}}{T}\right)_{\text{in}} + \dot{m}s_{\text{in}} - \left(\frac{\dot{Q}}{T}\right)_{\text{out}} - \dot{m}s_{\text{out}} + \dot{S}_{\text{gen}} = 0 \qquad (2.30)$$

In these equations T represents the temperature of the boundary at which heat transfer takes place. If the system is selected such that it includes the immediate surroundings, the boundary temperature becomes the temperature of the surroundings. Then one can use the surrounding ambient temperature in these equations. For the unsteady-flow process shown in Fig. 2.3, the entropy balance can be expressed as

$$\left(\frac{Q}{T}\right)_{\text{in}} + m_{\text{in}}s_{\text{in}} - \left(\frac{Q}{T}\right)_{\text{out}} - m_{\text{out}}s_{\text{out}} + S_{\text{gen}} = m_2 s_2 - m_1 s_1 \qquad (2.31)$$

In recent decades, much effort has been spent in minimizing the entropy generation (irreversibility) in thermodynamic systems and applications [8].

2.6 Exergy

The attempts to quantify the quality or "work potential" of energy in the light of the second law of thermodynamics has resulted in the definition of the exergy property.

Exergy analysis is a thermodynamic analysis technique based on the second law of thermodynamics that provides an alternative and illuminating means of assessing and comparing processes and systems rationally and meaningfully. In particular, exergy analysis yields efficiencies that provide a true measure of how nearly actual performance approaches the ideal, and identifies more clearly than energy analysis the causes and locations of thermodynamic losses and the impact of the built environment on the natural environment. Consequently, exergy analysis can assist in improving and optimizing designs. Various books have been written on exergy analysis and applications [5, 8–11].

Energy and exergy efficiencies are considered by many to be useful for the assessment of energy conversion and other systems and for efficiency improvement. By considering both of these efficiencies, the quality and quantity of the energy used to achieve a given objective is considered and the degree to which efficient and effective use of energy resources is achieved can be understood. Improving efficiencies of energy systems is an important challenge for meeting energy policy objectives. Reductions in energy use can assist in attaining energy security objectives. Also, efficient energy utilization and the introduction of renewable energy technologies can significantly help solve environmental issues. Increased energy efficiency benefits the environment by avoiding energy use and the corresponding resource consumption and pollution generation. From an economic as well as an environmental perspective, improved energy efficiency has great potential [2].

An engineer designing a system is often expected to aim for the highest reasonable technical efficiency at the lowest cost under the prevailing technical, economic, and legal conditions, and with regard to ethical, ecological, and social consequences. Exergy methods can assist in such activities and offer unique insights into possible improvements with special emphasis on environment and sustainability. Exergy analysis is a useful tool for addressing the environmental impact of energy resource utilization, and for furthering the goal of more efficient energy-resource use, for it enables the locations, types, and true magnitudes of losses to be determined. Also, exergy analysis reveals whether and by how much it is possible to design more efficient energy systems by reducing inefficiencies. We present exergy as a key tool for system and process analysis, design, and performance improvement.

2.6.1   What Is Exergy?

The useful work potential of a given amount of energy at a specified state is called *exergy*. It is also called the availability or available energy. The work potential of the energy contained in a system at a specified state, relative to a reference (dead) state, is simply the maximum useful work that can be obtained from the system [12].

A system is said to be in the dead-state when it is in thermodynamic equilibrium with its environment. At the dead-state, a system is at the temperature and pressure of its environment (in thermal and mechanical equilibrium), it has no kinetic or potential energy relative to the environment (zero velocity and zero elevation above a reference level), and it does not react with the environment (chemically inert). Also, there are no unbalanced magnetic, electrical, and surface tension effects between the system and its surroundings, if these are relevant to the situation at hand. The properties of a system at the dead-state are denoted by the subscript zero, for example, P_0, T_0, h_0, u_0, and s_0. Unless specified otherwise, the dead-state temperature and pressure are taken to be $T_0 = 25°C$ (77°F) and $P_0 = 1$ atm (101.325 kPa or 14.7 psia). A system has zero exergy at the dead-state.

The notion that a system must go to the dead-state at the end of the process to maximize work output can be explained as follows. If the system temperature at the final state is greater (or less) than the temperature of the environment it is in, we can always produce additional work by running a heat engine between these two temperature levels. If the final pressure is greater (or less) than the pressure of the environment, we can still obtain work by letting the system expand to the pressure of the environment. If the final velocity of the system is not zero, we can catch that extra kinetic energy by a turbine and convert it to rotating shaft work, and so on. No work can be produced from a system that is initially at the dead-state. The atmosphere around us contains a tremendous amount of energy. However, the atmosphere is in the dead-state, and the energy it contains has no work potential.

Therefore, we conclude that a system delivers the maximum possible work as it undergoes a reversible process from the specified initial state to the state of its environment, that is, the dead-state. It is important to realize that exergy does not represent the amount of work that a work-producing device will actually deliver upon installation. Rather, it represents the upper limit on the amount of work a device can deliver without violating any thermodynamic laws. There will always be a difference, large or small, between exergy and the actual work delivered by a device. This difference represents the available room that engineers have for improvement especially for greener buildings and more sustainable buildings per ASHRAE's Sustainability Roadmap.

Note that the exergy of a system at a specified state depends on the conditions of the environment (the dead-state) as well as the properties of the system. Therefore, exergy is a property of the system–environment combination and not of the system alone. Altering the environment is another way of increasing exergy, but it is definitely not an easy alternative.

The work potential or exergy of the kinetic energy of a system is equal to the kinetic energy itself because it can be entirely converted to work. Similarly, exergy of potential energy is equal to the potential energy itself. On the other hand, the internal energy and enthalpy of a system are not entirely available for work, and only part of the thermal energy of a system can be converted to work. In other words, the exergy of thermal energy is less than the magnitude of thermal energy.

2.6.2   Reversibility and Irreversibility

These two concepts are highly important in the analysis of thermodynamic processes and systems. The *reversibility* refers to a process during which both the system and its surroundings can be returned to their initial states. The irreversibility is associated with the destruction of exergy, and during an irreversible process, both the system and its surroundings cannot be returned to their initial states because of the irreversibilities occurring, for example, friction, heat rejection, electrical and mechanical effects, and the like.

2.6.3   Reversible Work and Exergy Destruction

The *reversible work* W_{rev} is defined as the maximum amount of useful work output or the minimum work input for a system undergoing a process between the specified initial and final states in a totally reversible manner.

Any difference between the reversible work W_{rev} and the actual work W_u is due to the irreversibilities present during the process, and this difference is called irreversibility or exergy destroyed. It is expressed as

$$X_{destroyed} = W_{rev,out} - W_{out} \quad \text{or} \quad X_{destroyed} = W_{in} - W_{rev,in} \tag{2.32}$$

Irreversibility is a positive quantity for all actual (irreversible) processes because $W_{rev} \geq W$ for work-producing devices and $W_{rev} \leq W$ for work-consuming devices.

Irreversibility can be viewed as the wasted work potential or the lost opportunity to do useful work. It represents the energy that could have been converted to work but was not. It is important to note that lost opportunities manifest themselves in environmental degradation and avoidable emissions. The smaller the irreversibility associated with a process, the greater the work that is produced (or the smaller the work that is consumed). The performance of a system can be improved by minimizing the irreversibility associated with it.

2.6.4   Exergy Change

A closed system, in general, may possess kinetic and potential energies, and in the absence of electric, magnetic, and surface tension effects, the total energy of a closed system is equal to the sum of its internal, kinetic, and potential energies. Noting that kinetic and potential energies themselves are forms of exergy, the exergy of a closed system of mass m is given by

$$X = (U - U_0) + P_0(V - V_0) - T_0(S - S_0) + m\frac{V^2}{2} + mgz \tag{2.33}$$

where the properties with the subscript zero represent those at the dead-state. On a unit mass basis, the closed system (or nonflow) exergy is expressed as

$$\phi = (u - u_0) + P_0(v - v_0) - T_0(s - s_0) + \frac{V^2}{V} + gz \tag{2.34}$$

The exergy change of a closed system during a process is simply the difference between the final and initial exergies of the system as follows:

$$\Delta X = U_2 - U_1 + P_0(V_2 - V_1) - T_0(S_2 - S_1) + m\frac{V_2^2 - V_1^2}{2} + mg(z_2 - z_1) \tag{2.35}$$

For stationary closed systems, the kinetic and potential energy terms drop out. The exergy of a flowing fluid is also called *flow* (or *stream*) *exergy*, and is given by

$$\psi = (h - h_0) - T_0(s - s_0) + \frac{V^2}{2} + gz \tag{2.36}$$

Then the *exergy change* of a fluid stream as it undergoes a process from state 1 to state 2 becomes

$$\Delta\psi = \psi_2 - \psi_1 = (h_2 - h_1) - T_0(s_2 - s_1) + \frac{V_2^2 - V_1^2}{2} + g(z_2 - z_1) \tag{2.37}$$

For fluid streams with negligible kinetic and potential energies, the kinetic and potential energy terms drop out.

Note that the exergy change of a closed system or a fluid stream represents the maximum amount of useful work that can be done (or the minimum amount of useful work that needs to be supplied if it is negative) as the system changes from state 1 to state 2 in a specified environment, and represents the reversible work W_{rev}. It is independent of the type of process executed, the kind of system used, and the nature of energy interactions with the surroundings. Also note that the exergy of a closed system cannot be negative, but the exergy of a flow stream can at pressures below the environment pressure P_0.

2.6.5 Exergy Transfer Mechanisms

Heat transfer Q at a location at thermodynamic temperature T is always accompanied by exergy transfer X_{heat} in the amount of

$$X_{\text{heat}} = \left(1 - \frac{T_0}{T}\right)Q \tag{2.38}$$

Exergy is the useful work potential, and the exergy transfer by work can simply be expressed as

$$X_{\text{work}} = W \tag{2.39}$$

and for boundary work,

$$X_{\text{work}} = W - W_{\text{surr}} = P_0(V_2 - V_1) \tag{2.40}$$

where P_0 is the atmospheric pressure and V_1 and V_2 are the initial and final volumes of the system.

Exergy transfer by mass is

$$X_{\text{mass}} = m\psi = m\left[(h - h_0) - T_0(s - s_0) + \frac{V^2}{2} + gz\right] \tag{2.41}$$

2.6.6   Exergy Balance

The nature of exergy is opposite to that of entropy in that exergy can be destroyed, but it cannot be created. Therefore, the exergy change of a system during a process is less than the exergy transfer by an amount equal to the exergy destroyed during the process within the system boundaries. Then the decrease of exergy principle can be expressed as

$$X_{\text{in}} - X_{\text{out}} - X_{\text{destroyed}} = \Delta X_{\text{system}} \tag{2.42}$$

In rate form,

$$\dot{X}_{\text{in}} - \dot{X}_{\text{out}} - \dot{X}_{\text{destroyed}} = \left(\frac{dX}{dt}\right)_{CV} \tag{2.43}$$

This relation is referred to as the exergy balance and can be stated as the exergy change of a system during a process is equal to the difference between the net exergy transfer through the system boundary and the exergy destroyed within the system boundaries as a result of irreversibilities. Exergy can be transferred to or from a system by heat, work, and mass.

Irreversibilities such as friction, mixing, chemical reactions, heat transfer through a finite temperature difference, unrestrained expansion, and nonquasi-equilibrium compression or expansion always generate entropy, and anything that generates entropy always destroys exergy. The exergy destroyed is proportional to the entropy generated, and is expressed as

$$X_{\text{destroyed}} = T_0 S_{\text{gen}} \tag{2.44}$$

Exergy destruction during a process can be determined from an exergy balance on the system or from the entropy generation using (2.44).

Fig. 2.4 A closed system
involving heat input Q_{in} and
boundary work output W_{out}

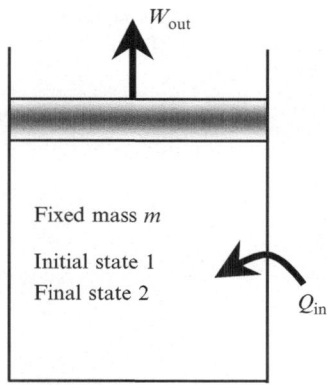

A closed system, in general, may possess kinetic and potential energies as the total energy involved. The exergy change of a closed system during a process is simply the exergy difference between the final state 2 and initial state 1 of the system. For a stationary closed system involving heat input Q_{in} and boundary work output W_{out} as shown in Fig. 2.4, mass, energy, entropy, and exergy balances can be expressed as

$$\text{Mass balance}: \quad m_1 = m_2 = \text{constant} \tag{2.45}$$

$$\text{Energy balance}: \quad Q_{in} - W_{out} = m(u_2 - u_1) \tag{2.46}$$

$$\text{Entropy balance}: \quad \frac{Q_{in}}{T_s} + S_{gen} = m(s_2 - s_1) \tag{2.47}$$

Exergy balance :

$$Q_{in}\left(1 - \frac{T_0}{T_s}\right) - [W_{out} - P_0(V_2 - V_1)] - X_{destroyed} = X_2 - X_1 \tag{2.48}$$

where u is internal energy, s is entropy, T_s is source temperature, T_0 is the dead-state (environment) temperature, S_{gen} is entropy generation, P_0 is the dead-state pressure, and V is volume. The exergy of a closed system is either positive or zero, and never becomes negative.

For a control volume involving a steady-flow process with heat input and power output as shown in Fig. 2.5, mass, energy, entropy, and exergy balances can be expressed as

$$\text{Mass balance}: \quad \dot{m}_1 = \dot{m}_2 \tag{2.49}$$

$$\text{Energy balance}: \quad \dot{m}_1 h_1 + \dot{Q}_{in} = \dot{m}_2 h_2 + \dot{W}_{out} \tag{2.50}$$

Fig. 2.5 A control volume
involving heat input and
power output

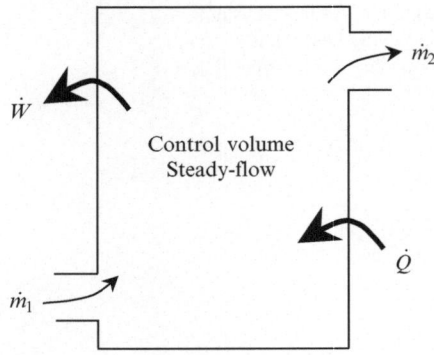

Control volume
Steady-flow

Entropy balance : $\dfrac{\dot{Q}_{in}}{T_s} + \dot{m}_1 s_1 + \dot{S}_{gen} = \dot{m}_2 s_2$ (2.51)

Exergy balance : $\dot{Q}_{in}\left(1 - \dfrac{T_0}{T_s}\right) + \dot{m}_1 \psi_1 = \dot{m}_2 \psi_2 + \dot{W}_{out} + \dot{X}_{destroyed}$ (2.52)

where specific exergy of a flowing fluid (i.e., flow exergy) is given by

$$\psi = h - h_0 - T_0(s - s_0)$$ (2.53)

In these equations, kinetic and potential energy changes are assumed to be negligible. Most control volumes encountered in practice such as turbines, compressors, heat exchangers, pipes, and ducts operate steadily, and thus they experience no changes in their mass, energy, entropy, and exergy contents as well as their volumes. The rate of exergy entering a steady-flow system in all forms (heat, work, mass transfer) must be equal to the amount of exergy leaving plus the exergy destroyed.

Chapter 3
Energy and Exergy Efficiencies

3.1 Introduction

Reductions in energy use can assist in attaining energy security objectives. Also, efficient energy utilization and the introduction of renewable energy technologies can significantly help solve environmental issues. Increased energy efficiency benefits the environment by avoiding energy use and the corresponding resource consumption and pollution generation. From an economic as well as an environmental perspective, improved energy efficiency has great potential [2].

An engineer designing a system is often expected to aim for the highest reasonable technical efficiency at the lowest cost under the prevailing technical, economic, and legal conditions, and with regard to ethical, ecological, and social consequences. Exergy methods can assist in such activities and offer unique insights into possible improvements with special emphasis on environment and sustainability. Exergy analysis is a useful tool for addressing the environmental impact of energy resource utilization, and for furthering the goal of more efficient energy-resource use, for it enables the locations, types, and true magnitudes of losses to be determined. Also, exergy analysis reveals whether and by how much it is possible to design more efficient energy systems by reducing inefficiencies.

Efficiency is a measure of the effectiveness and/or performance of a system. Although it may take different forms, depending on the application and purpose, it can generally be defined as

$$\eta = \frac{\text{Desired output}}{\text{Required input}} \tag{3.1}$$

The definition of energy efficiency is based on the first law of thermodynamics. It is denoted by η. It may take different forms and different names depending on the type of the system. It may be written as

$$\eta = \frac{\text{Energy output}}{\text{Energy input}} = \frac{E_{\text{out}}}{E_{\text{in}}} = 1 - \frac{E_{\text{loss}}}{E_{\text{in}}} \tag{3.2}$$

M. Kanoğlu et al., *Efficiency Evaluation of Energy Systems*,
SpringerBriefs in Energy, DOI 10.1007/978-1-4614-2242-6_3,
© Mehmet Kanoğlu, Yunus A. Çengel, İbrahim Dinçer 2012

where

$$E_{\text{in}} = E_{\text{output}} + E_{\text{loss}} \tag{3.3}$$

or in rate form,

$$\eta = \frac{\dot{E}_{\text{out}}}{\dot{E}_{\text{in}}} = 1 - \frac{\dot{E}_{\text{loss}}}{\dot{E}_{\text{in}}} \tag{3.4}$$

An alternative way of expressing energy efficiency is

$$\eta = \frac{\text{Energy recovered}}{\text{Energy expended}} = \frac{E_{\text{recovered}}}{E_{\text{expended}}} = 1 - \frac{E_{\text{loss}}}{E_{\text{expended}}} \tag{3.5}$$

where

$$E_{\text{expended}} = E_{\text{recovered}} + E_{\text{loss}} \tag{3.6}$$

or in rate form,

$$\eta = \frac{\dot{E}_{\text{recovered}}}{\dot{E}_{\text{expended}}} = 1 - \frac{\dot{E}_{\text{loss}}}{\dot{E}_{\text{expended}}} \tag{3.7}$$

Both (3.2) and (3.5) may be used to find the energy efficiency of a system but one may be more appropriate than the other depending on the system and application. They may turn out to be equivalent in some cases and different in others.

The definition of exergy efficiency is based on the second law of thermodynamics. It is also called second law efficiency or exergetic efficiency. Some sources also call it effectiveness. In this book we use exergy efficiency and second law efficiency interchangeably. Effectiveness is used with a different meaning for the performance of some devices.

Exergy efficiency may take different forms depending on the type of the system. It is denoted by η_{ex}, η_{II}, or ε. In this book, we use ε for the exergy efficiency symbol because it is easier to distinguish from the energy efficiency symbol. Exergy efficiency is generally expressed as

$$\varepsilon = \frac{\text{Exergy output}}{\text{Exergy input}} = \frac{X_{\text{out}}}{X_{\text{in}}} = 1 - \frac{X_{\text{dest}}}{X_{\text{in}}} \tag{3.8}$$

or in rate form,

$$\varepsilon = \frac{\dot{X}_{\text{out}}}{\dot{X}_{\text{in}}} = 1 - \frac{\dot{X}_{\text{dest}}}{\dot{X}_{\text{in}}} \tag{3.9}$$

where

$$\dot{X}_{in} = \dot{X}_{out} + \dot{X}_{dest} \tag{3.10}$$

Conceptually, second law efficiency is a measure of perfection. Thermodynamic perfection is reversibility. The second law dictates that no process can be better than a corresponding reversible process (in everyday terms, nothing can be more perfect than perfect), as that would be a violation of the second law. Therefore, reversible operation is the best possible mode of operation of a device, and thus it is natural that the second law efficiency be 1 or 100% for operations that involve no irreversibilities or imperfections. This sets the upper limit for second law efficiency, and current practice adheres to it.

What is essentially lacking is a fundamental definition for the lower limit of second law efficiency, which should be 0 or 0%. We should establish that the second law efficiency of a device or process that destroys the entire exergy it consumes is zero. The way to accomplish this is to change the general definition of second law efficiency from $\varepsilon = $ (Exergy output)/(Exergy input) to $\varepsilon = $ (Exergy recovered)/(Exergy expended). That is:

$$\varepsilon = \frac{X_{recovered}}{X_{expended}} = 1 - \frac{X_{dest}}{X_{expended}} \tag{3.11}$$

or in rate form,

$$\varepsilon = \frac{\dot{X}_{recovered}}{\dot{X}_{expended}} = 1 - \frac{\dot{X}_{dest}}{\dot{X}_{expended}} \tag{3.12}$$

where

$$\dot{X}_{expended} = \dot{X}_{recovered} + \dot{X}_{dest} \tag{3.13}$$

or alternately,

$$\varepsilon = \frac{\dot{X}_{delivered}}{\dot{X}_{consumed}} = 1 - \frac{\dot{X}_{dest}}{\dot{X}_{consumed}} \tag{3.14}$$

That is, the second-law efficiency of a device is the ratio of the exergy recovered (or delivered) by the device to the exergy expended (or consumed) by the device. This way, the second law efficiency of a device represents its ability to convert one form of exergy into another form (as from thermal to mechanical or vice versa). And, it refers to the resource and it puts the emphasis on the best utilization of a resource.

The difference between the two definitions may appear subtle, but it is fundamental. Here $\dot{X}_{expended}$ represents the portion of the exergy coming from the resource. It is the shaft work input in the case of a compressor, and the decrease in the exergy of steam (difference between exergy values at inlet and outlet) in the case of a steam turbine. Exergy recovered is the portion of the expended exergy that is retained as exergy, the portion that is saved from destruction within the system during the process.

Both (3.8) and (3.11) may be used to find the exergy efficiency of a system. In this book, we provide exergy efficiency formulations based on both approaches. However, for the reasons explained above, we recommend using the exergy recovered/exergy expended approach [(3.11)].

It should be mentioned that Cornelissen et al. [13] and Kotas [5] provide an exergy efficiency relation as the ratio of the desired exergy output to the exergy used. They call this rational efficiency. Here, exergy output is all exergy transfer from the system, plus any by-product that is produced by the system, whereas exergy used is the required exergy input for the process to be performed. This is similar to the exergy efficiency definition as the ratio of product to the fuel where the fuel represents the resources expended to generate the product [5, 13].

$$\varepsilon = \frac{\text{Product}}{\text{Fuel}} \tag{3.15}$$

Here, both product and fuel must be expressed in exergy terms.

The second law efficiency relation given by (3.11) is essentially the same as the rational efficiency definition of Cornelissen et al. [13] and Kotas [5]. However, most researchers utilize (3.8) for calculating exergy efficiencies and (3.11) is rarely used.

3.2   Efficiencies of Cyclic Devices

A heat engine is a device that converts heat to work. A steam power plant, a gas-turbine power plant, an internal combustion engine, a solar thermal plant, and a geothermal power plant are some familiar examples. Consider a heat engine as shown in Fig. 3.1. The high-temperature resource at T_H supplies heat to the heat engine at a rate of $\dot{Q}_H$. Some of this heat is converted to work $\dot{W}_{\text{out}}$ and the remaining heat $\dot{Q}_L$ is rejected to a low-temperature medium at T_L. The energy efficiency of this cycle is called thermal efficiency, and is expressed as the ratio of work produced to the heat supplied:

$$\eta_{\text{th}} = \frac{W_{\text{out}}}{Q_H} = 1 - \frac{Q_L}{Q_H} \tag{3.16}$$

Now, we consider a refrigeration or heat pump as shown in Fig. 3.2. A household refrigerator and an air-conditioning system used for cooling and heating are some familiar examples of such devices. Here heat at the rate of $\dot{Q}_L$ is absorbed from the low-temperature reservoir at T_L and heat at the rate of $\dot{Q}_H$ is rejected to the

Fig. 3.1 Schematic of a basic heat engine

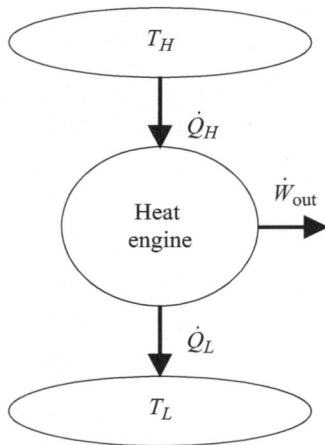

high-temperature reservoir at T_H. A power input $\dot{W}_{in}$ is needed for the operation of the cycle. The cycle is called a refrigerator if the purpose is to keep the low-temperature space at T_L and it is called a heat pump if the purpose of the cycle is to keep the high-temperature medium at T_H.

Because the desired output is different for a refrigerator and heat pump, their efficiencies are defined differently. If we utilize the general definition of efficiency, "desired output/required input," in this case, the performance of a refrigerator and a heat pump can be expressed by their coefficient of performance (COP):

$$COP_R = \frac{\dot{Q}_L}{\dot{W}_{in}} = \frac{\dot{Q}_L}{\dot{Q}_H - \dot{Q}_L} \tag{3.17}$$

$$COP_{HP} = \frac{\dot{Q}_H}{\dot{W}_{in}} = \frac{\dot{Q}_H}{\dot{Q}_H - \dot{Q}_L} \tag{3.18}$$

We are careful not to name this as efficiency because the COP values may be lower or greater than unity.

A heat engine that consists of all reversible processes is called a reversible heat engine or a Carnot heat engine. The thermal efficiency of a Carnot heat engine may be expressed by the temperatures of two reservoirs with which the heat engine exchanges heat (Fig. 3.1):

$$\eta_{th,\,rev} = 1 - \frac{T_L}{T_H} \tag{3.19}$$

where T_H is the source temperature and T_L is the sink temperature where heat is rejected (i.e., lake, ambient air, etc.). This is the maximum thermal efficiency a heat engine operating between two reservoirs at T_H and T_L can have.

Fig. 3.2 Schematic of a basic refrigeration or heat pump

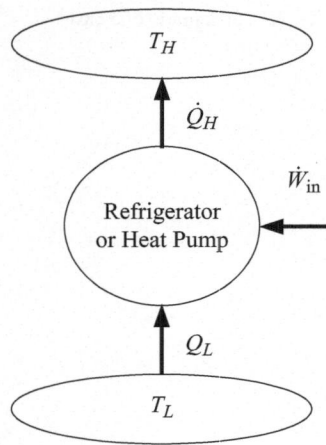

Because all processes in a Carnot cycle are reversible, the cycle can be reversed. In this case we obtain a reversed Carnot cycle. A refrigerator or heat pump operating on a reversed Carnot cycle (Fig. 3.2) would have the maximum COP values at the given temperature limits T_L and T_H, and they are expressible as

$$\text{COP}_{R,\text{rev}} = \frac{T_L}{T_H - T_L} \tag{3.20}$$

$$\text{COP}_{HP,\text{rev}} = \frac{T_H}{T_H - T_L} \tag{3.21}$$

Consider two heat engines, both having a thermal efficiency of 30%. One of the engines (engine A) receives heat from a source at 600 K, and the other one (engine B) from a source at 1,000 K. Both engines reject heat to a medium at 300 K. At first glance, both engines seem to be performing equally well. When we take a second look at these engines in light of the second law of thermodynamics, however, we see a totally different picture. These engines, at best, can perform as reversible engines, in which case their efficiencies in terms of the Carnot cycle become

$$\eta_{\text{th,rev},A} = \left(1 - \frac{T_0}{T_{\text{source}}}\right)_A = 1 - \frac{300\,\text{K}}{600\,\text{K}} = 0.5 \text{ or } 50\%$$

$$\eta_{\text{th,rev},B} = \left(1 - \frac{T_0}{T_{\text{source}}}\right)_B = 1 - \frac{300\,\text{K}}{1000\,\text{K}} = 0.7 \text{ or } 70\%$$

Engine A has a 50% useful work potential relative to the heat provided to it; engine B has 70%. Now it is becoming apparent that engine B has a greater work potential made available to it and thus should do a lot better than engine A. Therefore, we can say that engine B is performing poorly relative to engine A even though both have the same thermal efficiency.

It is obvious from this example that first law efficiency alone is not a realistic measure of performance of engineering devices. To overcome this deficiency, we define an exergy efficiency (or second law efficiency) for heat engines as the ratio of the actual thermal efficiency to the maximum possible (reversible) thermal efficiency under the same conditions:

$$\varepsilon = \frac{\eta_{th}}{\eta_{th,\,rev}} \tag{3.22}$$

Based on this definition, the exergy efficiencies of the two heat engines discussed above become

$$\varepsilon_A = \frac{0.30}{0.50} = 0.60 \text{ or } 60\%$$
$$\varepsilon_B = \frac{0.30}{0.70} = 0.43 \text{ or } 43\%$$

That is, engine A is converting 60% of the available work potential to useful work. This ratio is only 43% for engine B.

The second-law efficiency can be expressed for work-producing devices such as a turbine as the ratio of the useful work output to the maximum possible (reversible) work output:

$$\varepsilon = \frac{\dot{W}_{out}}{\dot{W}_{rev,\,out}} \tag{3.23}$$

This definition is more general because it can be applied to processes (in turbines, piston–cylinder devices, etc.) as well as to cycles. Note that the exergy efficiency cannot exceed 100%. We can also define an exergy efficiency for work-consuming noncyclic (such as compressors) and cyclic (such as refrigerators) devices as the ratio of the minimum (reversible) work input to the useful work input:

$$\varepsilon = \frac{\dot{W}_{rev,\,in}}{\dot{W}_{in}} \tag{3.24}$$

For cyclic devices such as refrigerators and heat pumps, it can also be expressed in terms of the coefficients of performance as

$$\varepsilon = \frac{COP}{COP_{rev}} \tag{3.25}$$

In the above relations, the reversible work should be determined by using the same initial and final states as in the actual process.

Example 3.1 A geothermal power plant uses geothermal liquid water at 160°C at a rate of 440 kg/s as the heat source, and produces 15 MW of net power in an environment at 25°C. Determine the thermal efficiency, the exergy efficiency, and the total rate of exergy destroyed in this power plant.

Solution The properties of geothermal water at the inlet of the plant and at the dead-state are obtained from steam tables to be

$$T_1 = 160°C, \text{ liquid} \longrightarrow h_1 = 675.47 \text{ kJ/kg}, s_1 = 1.9426 \text{ kJ/kg.K}$$
$$T_0 = 25°C, \ P_0 = 1 \text{ atm} \longrightarrow h_0 = 104.83 \text{ kJ/kg}, s_0 = 0.36723 \text{ kJ/kg.K}$$

The energy of geothermal water may be taken to be the maximum heat that can be extracted from the geothermal water, and this may be expressed as the enthalpy difference between the state of geothermal water and the dead-state:

$$\dot{E}_{in} = \dot{m}(h_1 - h_0) = (440 \text{ kg/s})[(675.47 - 104.83) \text{ kJ/kg}] = 251,080 \text{ kW}$$

The exergy of geothermal water is

$$\begin{aligned} \dot{X}_{in} &= \dot{m}[(h_1 - h_0) - T_0(s_1 - s_0)] \\ &= (440 \text{ kg/s})[(675.47 - 104.83) \text{ kJ/kg} + 0 \\ &\quad - (25 + 273 \text{ K})(1.9426 - 0.36723) \text{ kJ/kg.K}] = 44,525 \text{ kW} \end{aligned}$$

The thermal efficiency of the power plant is

$$\eta_{th} = \frac{\dot{W}_{net, \ out}}{\dot{E}_{in}} = \frac{15,000 \text{ kW}}{251,080 \text{ kW}} = 0.0597 \text{ or } 6.0\%$$

The exergy efficiency of the plant is the ratio of power produced to the exergy input to the plant:

$$\varepsilon = \frac{\dot{W}_{net, \ out}}{\dot{X}_{in}} = \frac{15,000 \text{ kW}}{44,525 \text{ kW}} = 0.337 \text{ or } 33.7\%$$

The exergy destroyed in this power plant is determined from an exergy balance on the entire power plant to be

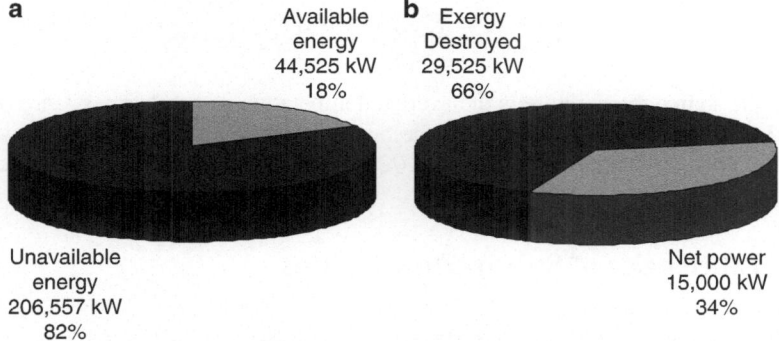

Fig. 3.3 (a) Only 18% of the energy of geothermal water is available for converting to power. (b) Only 34% of the exergy of geothermal water is converted to power and the remaining 66% is lost

$$\dot{X}_{\text{in}} - \dot{W}_{\text{net, out}} - \dot{X}_{\text{dest}} = 0$$
$$44,525 - 15,000 - \dot{X}_{\text{dest}} = 0 \longrightarrow \dot{X}_{\text{dest}} = 29,525\,\text{kW}$$

The results of this example are illustrated in Fig. 3.3a, b. The exergy of geothermal water constitutes only 18% of its energy, due to its well temperature. The remaining 82% is not available for useful work and it cannot be converted to power even by a reversible heat engine. Only 34% of exergy entering the plant is converted to power and the remaining 66% is lost. In geothermal power plants, the used geothermal water typically leaves the power plant at a temperature much greater than the environment temperature and this water is reinjected back to the ground. The total exergy destroyed (29,525 kW) includes the exergy of this reinjected brine.

In a typical binary-type geothermal power plant, geothermal water would be reinjected back to the ground at about 90°C. This water can be used in a district heating system. Assuming that geothermal water leaves the district at 70°C with a drop of 20°C during the heat supply, the rate of heat that could be used in the district system would be

$$\dot{Q}_{\text{heat}} = \dot{m}c\Delta T = (440\,\text{kg/s})(4.18\,\text{kJ/kg} \cdot^{\circ}\text{C})(20^{\circ}\text{C}) = 36,780\,\text{kW}$$

where c is the specific heat of water. This 36,780 kW heating is in addition to the 15,000 kW power generated. The energy efficiency of this cogeneration system would be $(15,000 + 36,780)/251,080 = 0.206$ or 20.6%. The energy efficiency increases from 6.0% to 20.6% as a result of incorporating a district heating system into the power plant.

The exergy of heat supplied to the district system is simply the heat supplied times the Carnot efficiency, which is determined as

$$\dot{X}_{\text{heat}} = \dot{Q}_{\text{heat}} \left(1 - \frac{T_0}{T_{\text{source}}} \right) = (36{,}780 \text{ kW}) \left(1 - \frac{278 \text{ K}}{353 \text{ K}} \right) = 7814 \text{ kW}$$

where the source temperature is the average temperature of geothermal water (80°C = 353 K) when supplying heat, and the environment temperature is taken as 5°C (278 K). This corresponds to 26.5% (7,814/29,525 = 0.265) of the exergy destruction.

3.3   Efficiencies of Steady-Flow Devices

In this section we provide various efficiencies used for steady-flow devices such as turbines, compressors, pumps, nozzles, diffusers, and heat exchangers. A power plant or a refrigeration system consists of a number of these steady-flow devices, and improving the performance of these devices would improve the performance of the entire plant or system.

3.3.1   Turbine

A fluid is expanded in a turbine to produce power. Steam and gas turbines are considered here. Turbines are normally well insulated so that their operation can be assumed to be adiabatic. The performance of an adiabatic turbine is usually expressed by isentropic (adiabatic) efficiency.

Consider a turbine with inlet state 1 with temperature T_1 and pressure P_1 and an exit state 2 with temperature T_2 (or steam quality) and pressure P_2. The power output from this compressor would be maximum if the fluid were expanded reversibly and adiabatically (i.e., isentropically) between the given initial state and given exit pressure. The isentropic efficiency is then the ratio of actual power to the isentropic power:

$$\eta_{isen, \text{ turbine}} = \frac{\dot{W}_{\text{actual}}}{\dot{W}_{\text{isentropic}}} = \frac{\dot{m}(h_1 - h_2)}{\dot{m}(h_1 - h_{2s})} \tag{3.26}$$

where $\dot{m}$ is the mass flow rate of fluid (kg/s) and h_s is the enthalpy of the fluid at the turbine outlet if the process were isentropic. This enthalpy may be obtained from exit pressure and exit entropy (equal to inlet entropy). Kinetic and potential energy changes are neglected.

Exergy efficiency of an adiabatic turbine may be determined from an "exergy recovered/exergy expended" approach. In this case, the exergy resource is steam, and exergy expended is the exergy supplied to steam to turbine, which is the decrease in the exergy of steam as it passes through the turbine. Exergy recovered is the shaft work. Taking state 1 as the inlet and state 2 as the outlet, the second law efficiency is expressed as

$$\varepsilon_{\text{turbine}-1} = \frac{\dot{X}_{\text{recovered}}}{\dot{X}_{\text{expended}}} = \frac{\dot{W}_{\text{out}}}{\dot{X}_1 - \dot{X}_2} = \frac{\dot{W}_{\text{out}}}{\dot{W}_{\text{rev}}} = \frac{\dot{m}(h_1 - h_2)}{\dot{m}[(h_1 - h_2 - T_0(s_1 - s_2)]} \quad (3.27)$$

$$\text{or} \quad \varepsilon_{\text{turbine}-1} = 1 - \frac{\dot{X}_{\text{dest}}}{\dot{X}_{\text{expended}}} = 1 - \frac{\dot{X}_{\text{dest}}}{\dot{X}_1 - \dot{X}_2} = 1 - \frac{\dot{W}_{\text{rev}} - \dot{W}_{\text{out}}}{\dot{X}_1 - \dot{X}_2} \quad (3.28)$$

The exergy efficiency definition based on the "exergy out/exergy in" approach is

$$\varepsilon_{\text{turbine}-2} = \frac{\dot{X}_{\text{out}}}{\dot{X}_{\text{in}}} = \frac{\dot{W}_{\text{out}} + \dot{X}_2}{\dot{X}_1} \neq \frac{\dot{W}_{\text{out}}}{\dot{W}_{\text{rev}}} \quad (3.29)$$

A third definition only assumes power output as the product and inlet exergy as the input:

$$\varepsilon_{\text{turbine}-3} = \frac{\dot{W}_{\text{out}}}{\dot{X}_1} = 1 - \frac{\dot{X}_{\text{dest}}}{\dot{X}_1} = 1 - \frac{\dot{W}_{\text{rev}} - \dot{W}_{\text{out}}}{\dot{X}_1} \quad (3.30)$$

Note that the first definition [(3.27)] is consistent with the general definition for the second law efficiency of work-producing devices (the ratio of actual work to reversible work), but the second and third definitions (3.29) and (3.30) are not. Also, the first definition satisfies both bounding conditions for the second law efficiency: it is 100% when actual work equals reversible work, and 0% when actual work is zero (and thus the entire expended exergy is destroyed).

It should be noted that isentropic efficiency and second law efficiency are different definitions. In isentropic efficiency, an ideal isentropic process between the actual initial state and an assumed hypothetical exit state is used whereas in exergy efficiency, an ideal reversible process between the actual inlet state and actual exit state is used. Consequently, close but different values for isentropic and exergy efficiencies are obtained. Some consequences of isentropic efficiency versus exergy efficiency for an adiabatic turbine are discussed in [14].

Example 3.2 Consider an adiabatic steam turbine with the following inlet and exit states: $P_1 = 10,000$ kPa, $T_1 = 500°C$, $P_2 = 10$ kPa, and $x_2 = 0.95$. Taking the dead-state temperature of steam as saturated liquid at $25°C$, determine the isentropic efficiency and exergy efficiency based on different approaches.

Solution The various efficiencies are determined from (3.26, 3.27, 3.29), and (3.30) to be

$$\eta_{\text{isen, turbine}} = 0.742, \quad \varepsilon_{\text{turbine}-1} = 0.812, \quad \varepsilon_{\text{turbine}-2} = 0.831, \quad \varepsilon_{\text{turbine}-3} = 0.729$$

That is, the second law efficiency is 72.9% based on (3.29) and it is 81.2% based on (3.27). In (3.29) and (3.30), the exergy of the steam at the turbine exit is part of the exergy destroyed by the turbine. However, the turbine should not be held responsible for the exergy it did not destroy as part of the processes associated

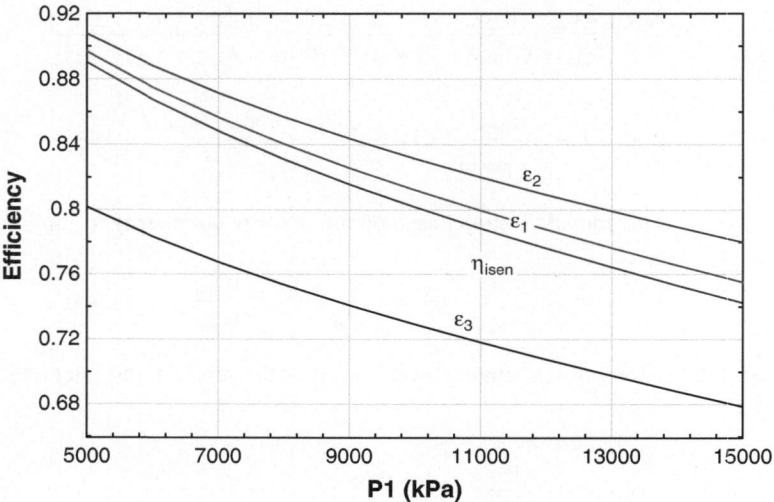

Fig. 3.4 Effect of turbine inlet pressure on the isentropic and second law efficiencies

with power production. With the first definition, the difference between the exergies of the inlet and exit steams is used for the exergy expended in the system.

The effect of turbine inlet pressure on isentropic efficiency [(3.26)] and three forms of exergy efficiencies [(3.27, 3.29, 3.30)] is investigated (Fig. 3.4). The efficiencies based on four definitions are considerably different. However, isentropic efficiency and second law efficiency by (3.27) are more appropriate forms. Interestingly, values of these two efficiencies are close to each other.

3.3.2   Compressor

A compressor is used to increase the pressure of a gas. A power input is needed for this compression process. The performance of an adiabatic compressor is usually expressed by isentropic (adiabatic) efficiency.

Consider an adiabatic compressor with inlet state 1 and an exit state 2. The power input to this compressor would be minimum if the gas were compressed reversibly and adiabatically (i.e., isentropically) between the given initial state and given exit pressure. The isentropic efficiency is then the ratio of the isentropic power to the actual power:

$$\eta_{isen, \text{ compressor}} = \frac{\dot{W}_{\text{isentropic}}}{\dot{W}_{\text{actual}}} = \frac{\dot{m}(h_{2s} - h_1)}{\dot{m}(h_2 - h_1)} \tag{3.31}$$

where $\dot{m}$ is the mass flow rate of the gas and h_s is the enthalpy of the fluid at the compressor outlet if the process were isentropic. This enthalpy may be obtained from exit pressure and exit entropy (equal to inlet entropy). Kinetic and potential energy changes are neglected. If the gas may be modeled as an ideal gas with constant specific heats, the isentropic power input is determined from

$$\dot{W}_{\text{isentropic}} = \dot{m}\frac{kR(T_2 - T_1)}{k - 1} = \dot{m}\frac{kRT_1}{k - 1}\left[\left(\frac{P_2}{P_1}\right)^{(k-1)/k} - 1\right] \qquad (3.32)$$

where k is the specific heat ratio ($k = c_p/c_v$). Its value is 1.4 for air at room temperature.

The gas is sometimes cooled as being compressed in a nonadiabatic compressor to reduce power input. This is because the power input is proportional to the specific volume of the gas and cooling the gas decreases its specific volume. The isentropic efficiency cannot be used in such nonadiabatic compressors. Instead, an isothermal efficiency may be defined as

$$\eta_{isothermal, \text{ compressor}} = \frac{\dot{W}_{\text{isothermal}}}{\dot{W}_{\text{actual}}} \qquad (3.33)$$

where the power input for the reversible isothermal case is given for an ideal gas with constant specific heats to be

$$\dot{W}_{\text{isothermal}} = \dot{m}RT_1 \ln\left(\frac{P_2}{P_1}\right) \qquad (3.34)$$

Here, R is the gas constant, T is the inlet temperature of the gas, and P_1 and P_2 are the pressures at the inlet and exit of the compressor, respectively. In some cases, a reversible polytropic process may be used as the ideal process for the actual compression. Then, a polytropic efficiency may be defined as

$$\eta_{polytropic, \text{ compressor}} = \frac{\dot{W}_{\text{polytropic}}}{\dot{W}_{\text{actual}}} \qquad (3.35)$$

where the power input for the reversible polytropic process is given for an ideal gas with constant specific heats to be

$$\dot{W}_{\text{polytropic}} = \dot{m}\frac{nR(T_2 - T_1)}{n - 1} = \dot{m}\frac{nRT_1}{n - 1}\left[\left(\frac{P_2}{P_1}\right)^{(n-1)/n} - 1\right] \qquad (3.36)$$

Here, n is the polytropic exponent and its value changes between 1 and specific heat ratio k.

The exergy efficiency of a compressor may be determined from an "exergy recovered/exergy expended" approach. Here the resource is shaft work input (which is exergy expended by the compressor), and exergy recovered is the exergy supplied to the working fluid of the compressor, which is the increase in the exergy of fluid as it passes through the compressor. Again taking state 1 as the inlet and state 2 as the outlet, the exergy efficiency is expressed as

$$\varepsilon_{compressor} = \frac{\dot{X}_{recovered}}{\dot{X}_{expended}} = \frac{\dot{X}_2 - \dot{X}_1}{\dot{W}_{in}} = \frac{\dot{W}_{rev}}{\dot{W}_{in}} \tag{3.37}$$

or

$$\varepsilon_{compressor} = 1 - \frac{\dot{X}_{dest}}{\dot{X}_{consumed}} = 1 - \frac{\dot{X}_{dest}}{\dot{W}_{in}} = 1 - \frac{\dot{W}_{in} - \dot{W}_{rev}}{\dot{W}_{in}} \tag{3.38}$$

The old definition based on the "exergy out/exergy in" approach is

$$\varepsilon_{compressor} = \frac{\dot{X}_{out}}{\dot{X}_{in}} = \frac{\dot{X}_2}{\dot{X}_1 + \dot{W}_{in}} \neq \frac{\dot{W}_{rev}}{\dot{W}_{in}} \tag{3.39}$$

Note that the definition of (3.37) is consistent with the general definition for the second law efficiency of work-consuming devices (the ratio of reversible work to actual work), but the definition of (3.39) is not. Also, the definition in (3.37) satisfies both bounding conditions for the second law efficiency: It is 100% when the exergy increase of the working fluid equals actual work input, and 0% when the fluid experiences no increase in exergy as it passes through the compressor (and thus the entire expended exergy is destroyed).

3.3.3 Pump

A pump is used to increase the pressure of a liquid. A power input is needed for this process. The liquid may be considered to be an incompressible fluid and the power input for the isentropic case may be determined from specific volume and pressure data. When the changes in potential and kinetic energies of a liquid are negligible, the isentropic efficiency of a pump is defined as

$$\eta_{isen,\,pump} = \frac{\dot{W}_{isentropic}}{\dot{W}_{actual}} = \frac{\dot{m}v(P_2 - P_1)}{\dot{m}(h_2 - h_1)} \tag{3.40}$$

where v is the specific volume of the liquid, and it is usually taken at the pump inlet.

One can also use (3.31) to determine isentropic efficiency of a pump. However, because the temperature change across a pump is small, it is difficult to get an accurate measurement of temperatures, and the corresponding enthalpy values are not dependable. For this reason, efficiency of a pump should be determined from the measurements of specific volume, pressures, and the actual power input. If the

Fig. 3.5 The cryogenic
turbine considered in
Example 3.3

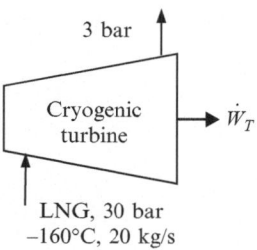

enthalpy value at the pump exit h_2 is needed, it is usually determined from a
knowledge of the pump isentropic efficiency or from the measurement of actual
power input.

The exergy efficiency of a pump can be determined from (3.37) as the reversible
power divided by the actual power where

$$\varepsilon_{pump} = \frac{\dot{X}_{recovered}}{\dot{X}_{expended}} = \frac{\dot{X}_2 - \dot{X}_1}{\dot{W}_{in}} = \frac{\dot{W}_{rev}}{\dot{W}_{in}} = \frac{\dot{m}[h_2 - h_2 - T_o(s_2 - s_1)]}{\dot{W}_{in}} \qquad (3.41)$$

We note again that it is difficult to get dependable values for enthalpy and
entropy due to the small temperature change across the pump. Therefore, it is
reasonable to assume that the reversible power input will be approximately equal
to the isentropic power input. Then, the definitions for the exergy efficiency and
isentropic efficiency become equal.

The operation of a hydraulic turbine is similar to that of a pump. Then the
efficiency of a hydraulic turbine may be expressed by modifying (3.40) as

$$\eta_{isen, \text{ hydraulic turbine}} = \frac{\dot{W}_{actual}}{\dot{W}_{isentropic}} = \frac{\dot{W}_{actual}}{\dot{m}v(P_1 - P_2)} \qquad (3.42)$$

This is also called hydraulic efficiency, and it may be considered as an isentropic
efficiency definition for a hydraulic turbine.

Example 3.3 Consider an adiabatic cryogenic turbine used in natural gas liquefaction plants (Fig. 3.5). Liquefied natural gas (LNG) enters a cryogenic turbine at
30 bar and $-160°C$ at a rate of 20 kg/s and leaves at 3 bar. If 115 kW power is
produced by the turbine, determine the efficiency of the turbine. Take the density of
LNG to be 423.8 kg/m^3.

Solution The maximum possible power that can be obtained from this turbine for
the given inlet and exit pressures can be determined from

$$\dot{W}_{isentropic} = \frac{\dot{m}}{\rho}(P_1 - P_2) = \frac{20 \, \text{kg/s}}{423.8 \, \text{kg/m}^3}(3000 - 300)\text{kPa} = 127.4 \, \text{kW}$$

Given the actual power, the efficiency of this cryogenic turbine becomes

$$\eta = \frac{\dot{W}_{actual}}{\dot{W}_{isentropic}} = \frac{115\,kW}{127.4\,kW} = 0.903 \text{ or } 90.3\%$$

3.3.4 Nozzle

A nozzle is essentially an adiabatic device because of the negligible heat transfer; it is used to accelerate a fluid. Therefore, the isentropic (i.e., reversible and adiabatic) process serves as a suitable model for nozzles. The *isentropic efficiency of a nozzle* is defined as the ratio of the actual kinetic energy of the fluid at the nozzle exit to the kinetic energy value at the exit of an isentropic nozzle for the same inlet state and exit pressure. That is,

$$\eta_{isen,\,nozzle} = \frac{KE_{exit,\,actual}}{KE_{exit,\,isentropic}} = \frac{V_2^2}{V_{2s}^2} \tag{3.43}$$

When the inlet velocity is negligible, the isentropic efficiency of the nozzle can be expressed in terms of enthalpies:

$$\eta_{isen,\,nozzle} = \frac{h_1 - h_2}{h_1 - h_{2s}} \tag{3.44}$$

A nozzle is built to convert the enthalpy of a fluid to kinetic energy, just like a turbine being built to convert the enthalpy of a fluid to shaft work. In a nozzle, the exergy recovered is the increase in the kinetic energy of the fluid and exergy expended is the exergy decrease of the fluid stream (but without taking into consideration the exit kinetic energy of the fluid stream, which corresponds to shaft work in a turbine). Then the exergy efficiency of a nozzle may be defined using an "exergy recovered/exergy expended" approach as

$$\varepsilon_{nozzle} = \frac{\dot{X}_{recovered}}{\dot{X}_{expended}} \tag{3.45}$$

and in a more explicit manner it becomes

$$\varepsilon_{nozzle} = \frac{\dot{X}_{recovered}}{\dot{X}_{expended}} = \frac{V_2^2/2 - V_1^2/2}{h_1 - h_2 - T_0(s_1 - s_2) + V_1^2/2}$$

$$\cong \frac{V_2^2/2}{h_1 - h_2 - T_0(s_1 - s_2)} \tag{3.46}$$

Note that expended exergy corresponds to the exit kinetic energy in a reversible process. When the nozzle is reversible and adiabatic, the exergy efficiency becomes 100%, as expected. For adiabatic nozzles, isentropic and exergy efficiencies are identical. For nonadiabatic nozzles, the denominator of (3.46) can be modified to include the exergy of the heat transferred.

3.3.5 Throttling Valve

A throttling valve is used to decrease the pressure of a fluid in a constant-enthalpy process. Because the energy of the mass at the inlet and exit of the valve are the same, it is not appropriate to define an efficiency based on the first law of thermodynamics. However, the adiabatic expansion process in a throttling valve is a highly irreversible process, and thus a second law efficiency may be defined.

Here the resource is the high pressure and temperature fluid, and exergy expended is the exergy supplied by the fluid, which is the decrease of the exergy of the fluid as it passes through the valve. There is no exergy recovered in any form, and thus exergy recovered is zero. Therefore:

$$\varepsilon_{\text{exp valve}} = \frac{\dot{X}_{\text{recovered}}}{\dot{X}_{\text{expended}}} = \frac{0}{\dot{X}_1 - \dot{X}_2} = 0 \tag{3.47}$$

$$\text{or} \quad \varepsilon_{\text{exp valve}} = 1 - \frac{\dot{X}_{\text{dest}}}{\dot{X}_{\text{expended}}} = 1 - \frac{\dot{X}_1 - \dot{X}_2}{\dot{X}_1 - \dot{X}_2} = 0 \tag{3.48}$$

This indicates that the second law efficiency of a throttling valve is always zero, which is expected because a throttling valve makes no use of the fluid exergy it expends. The commonly used definition for the exergy efficiency of a throttling valve is:

$$\varepsilon_{\text{exp valve}} = \frac{\dot{X}_{\text{out}}}{\dot{X}_{\text{in}}} = \frac{\dot{X}_2}{\dot{X}_1} \tag{3.49}$$

$$\text{or} \quad \varepsilon_{\text{exp valve}} = 1 - \frac{\dot{X}_{\text{dest}}}{\dot{X}_{\text{in}}} = 1 - \frac{\dot{X}_1 - \dot{X}_2}{\dot{X}_1} = \frac{\dot{X}_2}{\dot{X}_1} \tag{3.50}$$

Note that the current definition gives misleading results. A throttling valve that reduces the exergy rate of a fluid from 100 to 90 kW, for example, would have a second law efficiency of 90% according to the current definition, which is impressive. But this is unrealistic for such a wasteful device.

By similar reasoning, we can infer that the second law efficiency of electric transmission lines is zero inasmuch as the dissipated electric power is converted to heat which is rejected to the environment at environment temperature. Also, the second law efficiency of steam pipes losing heat to the environment is zero. Here the exergy of lost heat is zero because it is at the environment temperature, and no attempt is made to recover its exergy when the lost heat is at the outer surface temperature of the pipe.

3.3.6 Heat Exchanger

In a heat exchanger, two fluid streams exchange heat without mixing. When it comes to defining a second law or exergy efficiency for a heat exchanger, general disagreement is the rule rather than the exception. This is because there are several ways to define the exergy source. One way is to consider the hot fluid stream as the source of exergy and to disregard the cold fluid stream as a potential exergy source. Another way is to consider the exergy content of both fluid streams as the exergy source. Things are complicated even further when the cold fluid stream is below the environment temperature and the hot fluid stream is above so that the exergy of both fluid streams decreases during heat exchange.

When we deal with fluid streams above the environment temperature T_0, which is most often the case in practice, the cold fluid stream experiences an increase in its exergy content and none of the exergy of the incoming cold stream is expended. Therefore, we believe only the hot fluid stream should be considered as the exergy source in such cases, and the exergy increase of the cold fluid stream represents the exergy recovered. Then the exergy efficiency of a heat exchanger can be defined as

$$\varepsilon_{\text{heat exchanger}} = \frac{\dot{X}_{\text{recovered}}}{\dot{X}_{\text{expended}}} = \frac{\left(\dot{X}_{\text{out}} - \dot{X}_{\text{in}}\right)_{\text{cold}}}{\left(\dot{X}_{\text{in}} - \dot{X}_{\text{out}}\right)_{\text{hot}}} = 1 - \frac{\dot{X}_{\text{dest}}}{\left(\dot{X}_{\text{in}} - \dot{X}_{\text{out}}\right)_{\text{hot}}} \tag{3.51}$$

If the cold fluid is heated from T_1 to T_2 and the hot fluid is cooled from T_3 to T_4 at specified pressures, this expression may be written as

$$\varepsilon_{\text{heat exchanger}} = \frac{\left(\dot{X}_{\text{out}} - \dot{X}_{\text{in}}\right)_{\text{cold}}}{\left(\dot{X}_{\text{in}} - \dot{X}_{\text{out}}\right)_{\text{hot}}} = \frac{\dot{m}_{\text{cold}}[h_2 - h_1 - T_0(s_2 - s_1)]}{\dot{m}_{\text{hot}}[h_3 - h_4 - T_0(s_3 - s_4)]} \tag{3.52}$$

This relation will result in an exergy efficiency of 100% for a perfect counterflow heat exchanger where two identical fluids enter the heat exchanger at the same flow rates and the cold fluid is heated to the inlet temperature of the hot fluid while the hot fluid is cooled to the inlet temperature of the cold fluid. The exergy efficiency will be 0% for a heat exchanger that loses heat to its surroundings without transferring any heat to the cold fluid stream if the immediate surroundings of the heat exchanger are also taken as part of the system. Equation 3.51 is also valid if the heat exchanger is losing heat to the environment at temperature T_0 provided that the temperature gradient region between the heat exchanger and the environment is included in the analysis. If the heat exchanger is losing heat at a rate of $\dot{Q}_{loss}$ to a medium at T_R, the recovered exergy will also include the exergy stored in the medium at T_R as a result of this heat transfer. Then the exergy efficiency relation becomes

$$\varepsilon_{\text{heat exchanger}} = \frac{\dot{X}_{\text{recovered}}}{\dot{X}_{\text{expended}}} = \frac{\left(\dot{X}_{\text{out}} - \dot{X}_{\text{in}}\right)_{\text{cold}} + \dot{Q}_{\text{loss}}\left(1 - T_0/T_R\right)}{\left(\dot{X}_{\text{in}} - \dot{X}_{\text{out}}\right)_{\text{hot}}}$$

$$= 1 - \frac{\dot{X}_{\text{dest}}}{\left(\dot{X}_{\text{in}} - \dot{X}_{\text{out}}\right)_{\text{hot}}} \tag{3.53}$$

Special Case 1: Let us consider a heat exchanger where the hot fluid is cooled to a temperature above the environment temperature T_0, and the cold fluid remains below the environment temperature T_0 between the inlet and exit as it is heated. In this case, the exergies of both the hot and cold fluids will decrease and none of this exergy will be recovered. Therefore, in this case, it is more appropriate to take the "expended exergy" as the sum of the exergy decrease of the cold and hot fluids. This sum will be equal to the exergy destruction. As a result, the exergy efficiency in this case will be zero. That is,

$$\varepsilon_{\text{heatexchanger}} = \frac{\dot{X}_{\text{recovered}}}{\dot{X}_{\text{expended}}} = \frac{0}{\left(\dot{X}_{\text{in}} - \dot{X}_{\text{out}}\right)_{\text{hot}} + \left(\dot{X}_{\text{in}} - \dot{X}_{\text{out}}\right)_{\text{cold}}} = 0 \tag{3.54}$$

or

$$\varepsilon_{\text{heat exchanger}} = 1 - \frac{\dot{X}_{\text{dest}}}{\dot{X}_{\text{expended}}} = 1 - \frac{\left(\dot{X}_{\text{in}} - \dot{X}_{\text{out}}\right)_{\text{hot}} + \left(\dot{X}_{\text{in}} - \dot{X}_{\text{out}}\right)_{\text{cold}}}{\left(\dot{X}_{\text{in}} - \dot{X}_{\text{out}}\right)_{\text{hot}} + \left(\dot{X}_{\text{in}} - \dot{X}_{\text{out}}\right)_{\text{cold}}} = 0 \tag{3.55}$$

Special Case 2: What happens if the cold fluid is heated from a temperature T_1, which is below the environment temperature T_0 to a temperature T_2, which is greater than T_0 and the hot fluid is cooled from a temperature T_3, which is greater than T_0 to a temperature T_4, which is lower than T_0. In this case, the entire exergy contents of both the hot and cold fluid streams are expended, and thus the expended exergy is the sum of the initial exergies of the hot and cold fluid streams. But the hot fluid stream recovers part of this exergy as it is cooled to the subenvironment temperature and the cold fluid recovers part of its exergy as it is heated to the above environment temperature. Therefore, the total recovered exergy is the sum of the final exergies of the hot and cold fluid streams. Then the exergy efficiency in this case can be expressed as

$$\varepsilon_{\text{heatexchanger}} = \frac{\dot{X}_{\text{recovered}}}{\dot{X}_{\text{expended}}} = \frac{\dot{X}_{\text{out, cold}} + \dot{X}_{\text{out, hot}}}{\dot{X}_{\text{in, cold}} + \dot{X}_{\text{in, hot}}} = \frac{\dot{X}_2 + \dot{X}_4}{\dot{X}_1 + \dot{X}_3} \tag{3.56}$$

where

$$\dot{X}_1 = \dot{m}_{\text{cold}}[h_1 - h_0 - T_0(s_1 - s_0)] \quad \dot{X}_2 = \dot{m}_{\text{cold}}[h_2 - h_0 - T_0(s_2 - s_0)]$$
$$\dot{X}_3 = \dot{m}_{\text{hot}}[h_3 - h_0 - T_0(s_3 - s_0)] \quad \dot{X}_4 = \dot{m}_{\text{cold}}[h_4 - h_0 - T_0(s_4 - s_0)] \tag{3.57}$$

These relations are consistent with our general criteria that in the absence of any irreversibilities the exergy efficiency should be 100%. Indeed, the exergy efficiency for a perfect counterflow heat exchanger will be 100% if two identical fluid streams, one at $T_1 < T_0$ and the other at $T_2 > T_0$, enter the heat exchanger with identical mass flow rates because the cold fluid will be heated to T_2 and the hot fluid will be cooled to T_1.

To generalize, we can say that in the second law analysis of heat exchangers, all fluid streams that experience a decrease in their exergy content are to be considered in the evaluation of the expended exergy. Likewise, all fluid streams that experience an increase in their exergy content are to be considered in the evaluation of the recovered exergy. A fluid stream that crosses the dead-state is to be considered in the evaluation of both the expended exergy and recovered exergy. For such a fluid stream the expended exergy is the exergy at the inlet state, and the recovered exergy is the exergy at the exit state.

Effectiveness of a Heat Exchanger: The performance of heat exchangers is usually expressed by their effectiveness. It is then defined as

$$\eta_{\text{eff, heat exchanger}} = \frac{\dot{Q}_{\text{actual}}}{\dot{Q}_{\text{max}}} = \frac{\dot{Q}_{\text{actual}}}{(\dot{m}c_p)_{\text{min}}(T_{\text{hot,in}} - T_{\text{cold,in}})}$$
$$= \frac{(\dot{m}c_p \Delta T)_{\text{cold or hot}}}{(\dot{m}c_p)_{\text{min}}(T_{\text{hot, in}} - T_{\text{cold, in}})} \tag{3.58}$$

or using enthalpies

$$\eta_{\text{eff, heat exchanger}} = \frac{\dot{Q}_{\text{actual}}}{\dot{Q}_{\text{max}}} = \frac{(\dot{m}\Delta h)_{\text{cold or hot}}}{\dot{m}_{\text{min}}(h_{\text{hot, in}} - h_{\text{cold, in}})} \tag{3.59}$$

where $(\dot{m}c_p)_{\text{min}}$ is the smaller of the heat capacity rate between hot and cold fluids, and $\dot{m}_{\text{min}}$ is the smaller mass flow rate. The effectiveness of a heat exchanger is 100% when it transfers the maximum amount of heat transfer, and is 0% when it transfers no heat from the hot to the cold fluid.

Example 3.4 Calculate the exergy efficiency and effectiveness of a heat exchanger with the following data.

$$T_{1, \text{hot}} = 60°C, \; T_{2, \text{hot}} = 35°C, \; \dot{m}_{\text{hot}} = 1\,\text{kg/s}, \; c_{p, \text{hot}} = 4.18\,\text{kJ/kg} \cdot °\,C$$

$$T_{3, \text{cold}} = 5°C, \; T_{4, \text{cold}} = 20°C, \; c_{p, \text{cold}} = 4.18\,\text{kJ/kg} \cdot °\,C, \; T_0 = 25°C$$

Fig. 3.6 A cross-flow heat exchanger used to produce steam

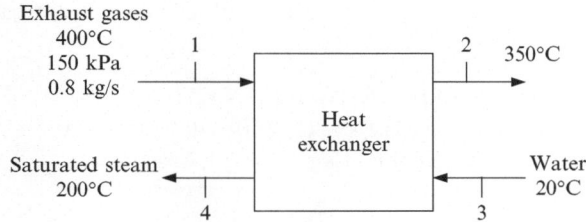

Exhaust gases
400°C
150 kPa
0.8 kg/s

Saturated steam
200°C

Heat exchanger

350°C

Water
20°C

Solution This heat exchanger operates as explained in Special Case 1. The exergy decreases of hot and cold fluids are determined to be 7.3 and 4.6 kW, respectively. The sum of these is 11.9 kW, and it is equal to the expended exergy. None of this expended exergy is recovered, therefore this is also equal to the exergy destruction, and the exergy efficiency is zero. The effectiveness is calculated from (3.58) as 0.456 or 45.6%.

Example 3.5 Calculate the exergy efficiency and effectiveness of a heat exchanger with the following data.

$$T_{1,\,\text{hot}} = 60^\circ\text{C}, \ T_{2,\,\text{hot}} = 15^\circ\text{C}, \ \dot{m}_{\text{hot}} = 1\,\text{kg/s}, \ c_{p,\,\text{hot}} = 4.18\,\text{kJ/kg} \cdot {}^\circ\text{C}$$

$$T_{3,\,\text{cold}} = 10^\circ\text{C}, \ T_{4,\,\text{cold}} = 35^\circ\text{C}, \ c_{p,\,\text{cold}} = 4.18\,\text{kJ/kg} \cdot {}^\circ\text{C}, \ T_0 = 25^\circ\text{C}$$

Solution This heat exchanger operates as explained in Special Case 2. Then, the exergy efficiency can be calculated using (3.56) to be 0.179 or 17.9%. The effectiveness is calculated from (3.58) as 0.90 or 90.0%.

Example 3.6 Hot exhaust gases leaving an internal combustion engine at 400°C and 150 kPa at a rate of 0.8 kg/s are to be used to produce saturated steam at 200°C in an insulated heat exchanger (Fig. 3.6). Water enters the heat exchanger at the ambient temperature of 20°C, and the exhaust gases leave the heat exchanger at 350°C. Determine the rate of steam production, the rate of exergy destruction in the heat exchanger, and the exergy efficiency of the heat exchanger.

Solution We denote the inlet and exit states of exhaust gases by (1) and (2) and that of the water by (3) and (4). The properties of water are obtained from the steam tables to be

$T_3 = 20^\circ\text{C}$, liquid $\longrightarrow h_3 = 83.91\,\text{kJ/kg}, s_3 = 0.29649\,\text{kJ/kg.K}$

$T_4 = 200^\circ\text{C}$, saturated vapor $\longrightarrow h_4 = 2792.0\,\text{kJ/kg}, s_4 = 6.4302\,\text{kJ/kg.K}$

An energy balance on the heat exchanger gives the rate of steam production:

$$\dot{m}_a h_1 + \dot{m}_w h_3 = \dot{m}_a h_2 + \dot{m}_w h_4$$
$$\dot{m}_a c_p (T_1 - T_2) = \dot{m}_w (h_4 - h_3)$$
$$(0.8\,\text{kg/s})(1.063\,\text{kJ/kg}^\circ\text{C})(400 - 350)^\circ\text{C} = \dot{m}_w (2792.0 - 83.91)\text{kJ/kg}$$
$$\dot{m}_w = 0.01570\,\text{kg/s}$$

The specific exergy changes of air and water streams as they flow in the heat exchanger are

$$
\begin{aligned}
\Delta\psi_a &= c_p(T_2 - T_1) - T_0(s_2 - s_1) \\
&= (1.063\,\text{kJ/kg.}^\circ\text{C})(350 - 400)^\circ\text{C} - (20 + 273\text{K})(-0.08206\,\text{kJ/kg.K}) \\
&= -29.106\,\text{kJ/kg}
\end{aligned}
$$

$$
\begin{aligned}
\Delta\psi_w &= (h_4 - h_3) - T_0(s_4 - s_3) \\
&= (2792.0 - 83.91)\text{kJ/kg} - (20 + 273\,\text{K})(6.4302 - 0.29649)\,\text{kJ/kg.K} \\
&= 910.913\,\text{kJ/kg}
\end{aligned}
$$

The exergy destruction is determined from an exergy balance as

$$
(\dot{m}_a\psi_1 + \dot{m}_w\psi_3) - (\dot{m}_a\psi_3 + \dot{m}_w\psi_4) - \dot{X}_{\text{destroyed}} = 0
$$

Rearranging and substituting,

$$
\begin{aligned}
\dot{X}_{\text{destroyed}} &= \dot{m}_a\Delta\psi_a + \dot{m}_w\Delta\psi_w \\
&= (0.8\,\text{kg/s})(-29.106\,\text{kJ/kg}) + (0.01570\,\text{kg/s})(910.913)\text{kJ/kg} \\
&= 8.98\,\text{kW}
\end{aligned}
$$

The exergy efficiency for a heat exchanger may be defined as the exergy increase of the cold fluid divided by the exergy decrease of the hot fluid. That is,

$$
\varepsilon = \frac{\dot{m}_w\Delta\psi_w}{-\dot{m}_a\Delta\psi_a} = \frac{(0.01570\,\text{kg/s})(910.913\,\text{kJ/kg})}{-(0.8\,\text{kg/s})(-29.106\,\text{kJ/kg})} = 0.614 \text{ or } 61.4\%
$$

The energy efficiency of this heat exchange process is 100% because it is assumed that all the energy given up by the exhaust gases are picked up by the water. The process is perfect from an energetic point of view whereas it is far from ideal from an exergetic perspective. The exergy destruction during this process is due to heat transfer across a finite temperature difference. This is illustrated in Fig. 3.7, which shows the exergy efficiency of the heat exchanger as a function of the exhaust gas inlet temperature for the same temperature drop of 50°C for the exhaust gases. As the average temperature difference between the exhaust gases and the water increases, the exergy efficiency decreases. In another words, the larger the temperature difference is, the larger the exergy destruction.

3.3.7   Mixing Chamber

Two fluid streams mix to produce a third fluid stream in a mixing chamber. When both incoming fluid streams are above the environment temperature, the exergy resource is the hot fluid, and the exergy expended is the exergy decrease of the hot fluid. The exergy recovered is the exergy increase of the cold fluid. Taking state 1 as the hot fluid inlet, state 2 as the cold fluid inlet, and state 3 as the common state of the mixture,

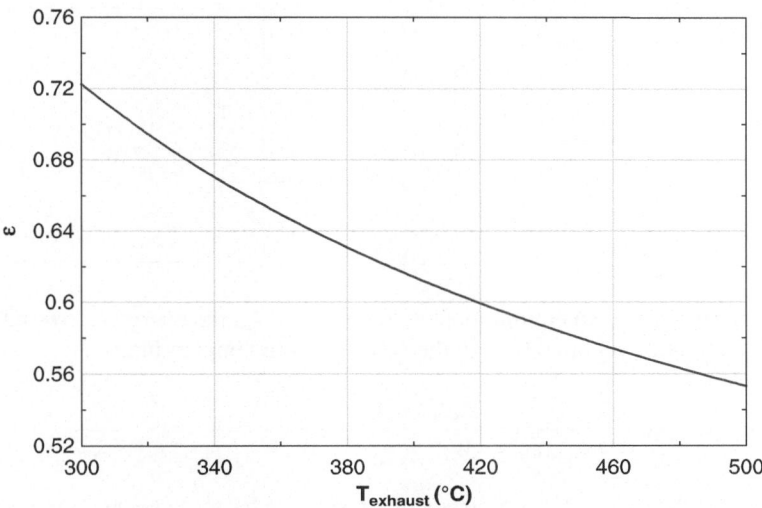

Fig. 3.7 The exergy efficiency of a heat exchanger as a function of exhaust gas inlet temperature

$$
\varepsilon_{\text{mixing chamber}} = \frac{\dot{X}_{\text{recovered}}}{\dot{X}_{\text{expended}}} = \frac{\left(\dot{X}_{\text{out}} - \dot{X}_{\text{in}}\right)_{\text{cold}}}{\left(\dot{X}_{\text{in}} - \dot{X}_{\text{out}}\right)_{\text{hot}}} = \frac{\dot{X}_3 - \dot{X}_2}{\dot{X}_1 - \dot{X}_3}
$$

$$
= \frac{\dot{m}_{\text{cold}}[h_3 - h_2 - T_0(s_3 - s_2)]}{\dot{m}_{\text{hot}}[h_1 - h_3 - T_0(s_1 - s_3)]} = 1 - \frac{\dot{X}_{\text{dest}}}{\left(\dot{X}_{\text{in}} - \dot{X}_{\text{out}}\right)_{\text{hot}}}
\tag{3.60}
$$

Noting that $\dot{m}_{\text{cold}}(h_3 - h_2) = \dot{m}_{\text{hot}}(h_1 - h_3)$, manipulating the last equality gives $\dot{X}_{\text{dest}} = T_0[\dot{m}_{\text{hot}}(s_1 - s_3) + \dot{m}_{\text{cold}}(s_2 - s_3)]$, which provides a check.

If the mixing chamber is losing heat at a rate of $\dot{Q}_{\text{loss}}$ to a medium at T_R, the recovered exergy will also include the exergy associated with heat transfer,

$$
\varepsilon_{\text{mixing chamber}} = \frac{\dot{m}_{\text{cold}}[h_3 - h_2 - T_0(s_3 - s_2)] + \dot{Q}_{\text{loss}}(1 - T_0/T_R)}{\dot{m}_{\text{hot}}[h_1 - h_3 - T_0(s_1 - s_3)]}
\tag{3.61}
$$

When the cold stream is below the environment temperature, arguments similar to those given above for heat exchangers can be given.

3.3.8 Electric Resistance Heating

Here the resource is the electrical energy in the grid, and the exergy expended is the exergy of electricity expended by the resistance heater. If the heater is indoors at

Fig. 3.8 An electric
resistance heater used to heat
a room

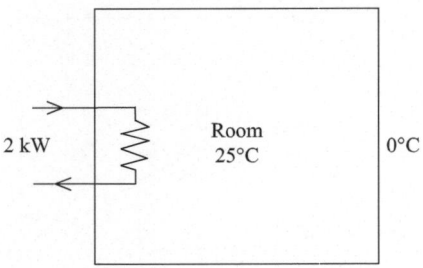

temperature T_{room} in an environment at temperature T_0, the exergy recovered is the
exergy content of supplied heat to the room at room temperature:

$$\varepsilon_{\text{electric heater}} = \frac{\dot{X}_{\text{recovered}}}{\dot{X}_{\text{expended}}} = \frac{\dot{X}_{\text{heat}}}{\dot{W}_e} = \frac{\dot{Q}_e(1 - T_0/T_{\text{room}})}{\dot{W}_e} = 1 - \frac{T_0}{T_{\text{room}}} \qquad (3.62)$$

Note that the second law efficiency of a resistance heater becomes zero when the
heater is outdoors.

Example 3.7 An electric resistance heater with a power consumption of 2.0 kW is
used to heat a room at 25°C when the outdoor temperature is 0°C (Fig. 3.8).
Determine energy and exergy efficiencies and the rate of exergy destroyed for
this process.

Solution For each unit of electric work consumed, the heater will supply the house
with 1 unit of heat. That is, the heater has a COP of 1. Also, the energy efficiency of
the heater is 100% because the energy output (heat supply to the room) and the
energy input (electric work consumed by the heater) are the same. At the specified
indoor and outdoor temperatures, a reversible heat pump would have a COP of

$$\text{COP}_{\text{HP, rev}} = \frac{1}{1 - T_L/T_H} = \frac{1}{1 - (273\,\text{K})/(298\,\text{K})} = 11.9$$

That is, it would supply the house with 11.9 units of heat (extracted from the cold
outside air) for each unit of electric energy it consumes (Fig. 3.9). The exergy
efficiency of this resistance heater is

$$\varepsilon = \frac{\text{COP}}{\text{COP}_{\text{HP, rev}}} = \frac{1}{11.9} = 0.084 \text{ or } 8.4\%$$

The minimum work requirement to the heater is determined from the COP
definition for a heat pump to be

$$\dot{W}_{\text{in, min}} = \frac{\dot{Q}_{\text{supplied}}}{\text{COP}_{\text{HP, rev}}} = \frac{2\,\text{kW}}{11.9} = 0.17\,\text{kW}$$

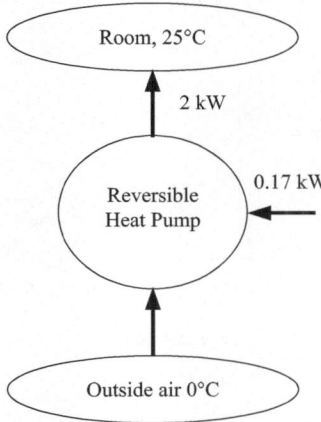

Fig. 3.9 A reversible heat pump consuming only 0.17 kW power while supplying 2 kW of heat to a room

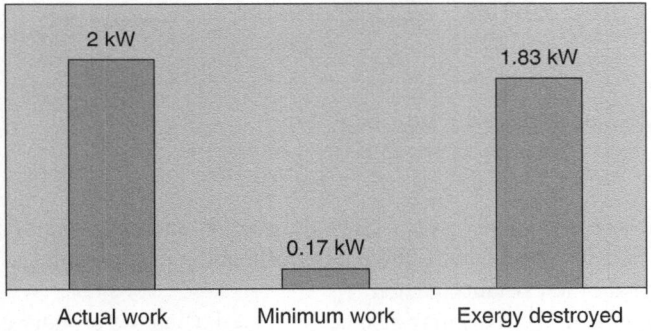

Fig. 3.10 Comparison of actual and minimum works with the exergy destroyed

That is, a reversible heat pump would consume only 0.17 kW of electrical energy to supply the room with 2 kW of heat. The exergy destroyed is the difference between the actual and minimum work input:

$$\dot{X}_{\text{destroyed}} = \dot{W}_{\text{in}} - \dot{W}_{\text{in, min}} = 2.0 - 0.17 = 1.83\,\text{kW}$$

The results of this example are illustrated in Figs. 3.10 and 3.11. The performance looks perfect with respect to energy efficiency but not so good from the point of view of exergy efficiency. About 92% of actual work input to the resistance heater is wasted during the operation of the resistance heater. There must be better methods of heating this room. Using a heat pump (preferably a ground-source one) or a natural

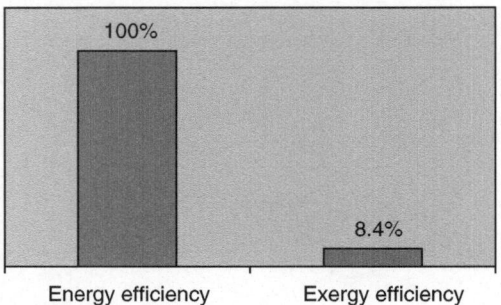

Fig. 3.11 Comparison of energy and exergy efficiencies

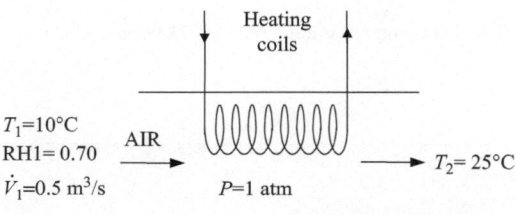

Fig. 3.12 Schematic of a simple heating process

gas furnace would involve lower exergy destruction and correspondingly greater exergy efficiencies even though the energy efficiency of a natural gas furnace is lower than that of a resistance heater.

Different heating systems may also be compared using the primary energy ratio (PER), which is the ratio of useful heat delivered to the primary energy input. Obviously, the higher the PER is, the more efficient the heating system. The PER for a heat pump is defined as $PER = \eta \times COP$ where η is the thermal efficiency with which the primary energy input is converted into work. For the resistance heater discussed in this example, the thermal efficiency η may be taken to be 0.40 if the electricity is produced from a natural gas-fueled steam power plant. Because the COP is 1, the PER becomes 0.40. A natural gas furnace with an efficiency of 0.80 (i.e., heat supplied over the heating value of the fuel) would have a PER value of 0.80. Furthermore, with a ground-source heat pump using electricity as the work input, the COP may be taken as 3 and with the same method of electricity production ($\eta = 0.40$), the PER becomes 1.2.

Example 3.8 In an air-conditioning process, air is heated by a heating coil in which hot water is flowing at an average temperature of 80°C. Using the values given in Fig. 3.12, determine the exergy destruction and the exergy efficiency for this process.

Solution The properties of air at various states (including dead-state, denoted by the subscript 0) are determined from software with built-in properties to be

$$v_1 = 0.810 \text{ m}^3/\text{kg}, \ h_0 = h_1 = 25.41 \text{ kJ/kg}, \ h_2 = 40.68 \text{ kJ/kg},$$

$$s_0 = s_1 = 5.701 \text{ kJ/kg} \cdot \text{K } s_2 = 5.754 \text{ kJ/kg} \cdot \text{K},$$

$$w_1 = w_2 = 0.00609 \text{ kg water/kg air}, \ RH_2 = 0.31$$

The dead-state temperature is taken to be the same as the inlet temperature of air. The mass flow rate of air and the rate of heat input are

$$\dot{m}_a = \frac{\dot{V}_a}{v_1} = 0.617 \text{ kg/s}$$

$$\dot{Q}_{in} = \dot{m}_a(h_2 - h_1) = 9.43 \text{ kW}$$

The exergies of the air stream at the inlet and exit become

$$\dot{X}_1 = 0 \quad \text{and} \quad \dot{X}_2 = \dot{m}_a[(h_2 - h_0) - T_0(s_2 - s_0)] = 0.267 \text{ kW}$$

The rates of exergy input and the exergy destroyed are

$$\dot{X}_{in} = \dot{Q}_{in}\left(1 - \frac{T_0}{T_{source}}\right) = 1.87 \text{ kW}$$

$$\dot{X}_{destroyed} = \dot{X}_{in} - \dot{X}_{out} = 1.87 - 0.267 = 1.60 \text{ kW}$$

where the temperature at which heat is transferred to the air stream is taken as the average temperature of water flowing in the heating coils (80°C). The exergy efficiency is

$$\varepsilon = \frac{\dot{X}_{out}}{\dot{X}_{in}} = \frac{0.267 \text{ kW}}{1.87 \text{ kW}} = 0.143 \text{ or } 14.3\%$$

About 86% of exergy input is destroyed due to irreversible heat transfer in the heating section. Air-conditioning processes typically involve high rates of exergy destructions as high-temperature (i.e., high-quality) heat or high-quality electricity is used to obtain a low-quality product. The irreversibilities can be minimized using lower quality energy sources and fewer irreversible processes. For example, if heat is supplied at an average temperature of 60°C instead of 80°C, the exergy destroyed would decrease from 1.60 to 1.15 kW and the exergy efficiency would increase from 14.3% to 18.8%. The exit temperature of air also affects the exergy efficiency. For example, if air is heated to 20°C instead of 25°C, the exergy efficiency would decrease from 14.3% to 10.1%. These two examples also show that the smaller the temperature difference between the heat source and the air being heated is, the larger the exergy efficiency.

Chapter 4
Energy Conversion Efficiencies

4.1 Conversion Efficiencies of Common Devices

Performance or efficiency, in general, is expressed in terms of the desired output and the required input as

$$\text{Efficiency} = \frac{\text{Desired output}}{\text{Required input}} \tag{4.1}$$

In this section, we provide definitions for energy conversion efficiency of some common devices.

4.1.1 Electric Resistance Heater

An electric resistance heater converts the electrical energy consumed by the device to the heat supplied to the space. Its conversion efficiency is then defined as

$$\eta = \frac{\dot{Q}}{\dot{W}_e} \tag{4.2}$$

where $\dot{Q}$ is the rate of heat supplied and $\dot{W}_e$ is the electrical power consumed.

What is the conversion efficiency of an electric resistance heater with a power rating of 2.4 kW when it is on? This heater consumes 2.4 kW electricity. The conservation of energy principle requires that all of this electricity be converted to heat, and 2.4 kW of heat will be supplied to the space to be heated. Therefore, its conversion efficiency is 100%. Work can be converted to heat 100% as the work is the highest quality form of energy and heat is the lowest quality one. On the other hand, the Kelvin–Plank statement of the second law of thermodynamics expresses that heat-to-work conversion must always be less than 100%.

M. Kanoğlu et al., *Efficiency Evaluation of Energy Systems*,
SpringerBriefs in Energy, DOI 10.1007/978-1-4614-2242-6_4,
© Mehmet Kanoğlu, Yunus A. Çengel, İbrahim Dinçer 2012

4.1.2   Electric Water Heater

The efficiency of an electric water heater is defined as the ratio of the energy delivered to the house by hot water to the electricity supplied to the water heater:

$$\eta = \frac{\dot{Q}_{\text{water}}}{\dot{W}_e} = \frac{\dot{m}_w(h_2 - h_1)}{\dot{W}_e} = \frac{\dot{m}_w c_p(T_2 - T_1)}{\dot{W}_e} \tag{4.3}$$

where $\dot{Q}_{\text{water}}$ is the rate of heat supplied to water (kW), $\dot{W}_e$ is the electrical power consumed (kW), $\dot{m}_w$ is the mass flow rate of water (kg/s), h_1 and h_2 are enthalpies of water at the inlet and outlet of the water heater (kJ/kg), T_1 and T_2 are temperatures of water at the inlet and outlet of the heater (°C), and c_p is the specific heat of water (kJ/kg· °C).

The efficiency of a conventional electric water heater is about 90%. You may find this confusing inasmuch as the heating elements of electric water heaters are resistance heaters, and the efficiency of all resistance heaters is 100% as they convert all the electrical energy they consume into thermal energy. The meaning of 90% conversion efficiency is that the heat losses from the hot-water tank to the surrounding air amount to 10% of the electrical energy consumed.

4.1.3   Natural Gas Water Heater

The efficiency of a natural gas water heater is defined as the ratio of the energy delivered to the house by hot water to the energy supplied to the heater by natural gas:

$$\eta = \frac{\dot{Q}_{\text{water}}}{\dot{m}_{\text{fuel}}\text{HV}} = \frac{\dot{m}_w(h_2 - h_1)}{\dot{m}_{\text{fuel}}\text{HV}} = \frac{\dot{m}_w c_p(T_2 - T_1)}{\dot{m}_{\text{fuel}}\text{HV}} \tag{4.4}$$

where HV is the heating value of the fuel (kJ/kg) and $\dot{m}_{\text{fuel}}$ is the consumption rate of fuel (kg/s).

4.1.4   Combustion Efficiency

The efficiency of equipment that involves the combustion of a fuel is based on the heating value of the fuel. The performance of combustion equipment can be characterized by *combustion efficiency*, defined as

$$\eta_{\text{combustion}} = \frac{\dot{Q}_{\text{released}}}{\dot{m}_{\text{fuel}}\text{HV}} \tag{4.5}$$

where $\dot{Q}_{\text{released}}$ is the rate of heat released during combustion.

A combustion efficiency of 100% indicates that the amount of heat released during a combustion process is equal to the heating value of the fuel. As an example, the combustion efficiency of a diesel automobile engine is about 98%. The 2% loss is mostly due to unburned fuel and carbon monoxide emission (CO is a fuel with a heating value as well as being a very undesirable pollutant) due to the very short time of combustion and nonhomogeneous air–fuel mixture. The combustion efficiency of gasoline engines is lower due to the smaller air–fuel mixture in the combustion chamber.

4.1.5 Heating Values

The *heating value* of the fuel is the amount of heat released when a unit amount of fuel at room temperature is completely burned and the combustion products are cooled to room temperature. Most fuels contain hydrogen, which forms water when burned, and the heating value of a fuel will be different, depending on whether the water in combustion products is in liquid or vapor form. The heating value is called the *lower heating value,* or LHV, when the water leaves as a vapor, and the *higher heating value,* or HHV, when the water in the combustion gases is completely condensed and thus the heat of vaporization is also recovered. The difference between these two heating values is equal to the product of the amount of water and the enthalpy of vaporization of water at room temperature. For example, the lower and higher heating values of gasoline are 44,000 kJ/kg and 47,300 kJ/kg, respectively. An efficiency definition should make it clear whether it is based on the higher or lower heating value of the fuel. Efficiencies of cars and jet engines are normally based on lower heating values because water normally leaves as a vapor in the exhaust gases, and it is not practical to try to recover the heat of vaporization.

4.1.6 Boiler Efficiency

A boiler is used to obtain steam by transferring the heat of a burning fuel. A boiler can also be used to heat water as in a natural gas water heater or to heat industrial oil or another fluid. When steam is obtained, the efficiency of a boiler can be defined as

$$\eta = \frac{\dot{Q}_{\text{useful}}}{\dot{m}_{\text{fuel}}\text{HV}} = \frac{\dot{m}_{\text{steam}}(h_2 - h_1)}{\dot{m}_{\text{fuel}}\text{HV}} \tag{4.6}$$

where $\dot{Q}_{\text{useful}}$ is the rate of useful heat supplied to the water, h_1 is the enthalpy of water at the boiler inlet, and h_2 is the enthalpy of steam at the boiler outlet. The steam usually leaves the boiler as a saturated vapor. In this case h_2 is the enthalpy of the saturated vapor at the boiler pressure. If the water is saturated liquid at the boiler inlet and saturated vapor at the boiler exit, the boiler efficiency may be determined from

$$\eta = \frac{\dot{Q}_{\text{useful}}}{\dot{m}_{\text{fuel}}\text{HV}} = \frac{\dot{m}_{\text{steam}}h_{fg}}{\dot{m}_{\text{fuel}}\text{HV}} \tag{4.7}$$

where h_{fg} is the enthalpy of vaporization of water at the given boiler pressure.

You may have realized that some manufacturers list the efficiencies of their boilers to be greater than 100%. Is this possible? See the following example.

Example 4.1 A natural gas fueled boiler is used to produce saturated steam at 5 bar pressure. Water enters the boiler at 70°C at a rate of 58 L/min. It is measured that 500 m^3 natural gas entering at 25°C and 101 kPa are consumed during a two-hour test. Calculate the efficiency of the boiler based on the lower and higher heating values of natural gas.

Solution The properties of water at the inlet and exit of the boiler are obtained from steam tables to be

$$T_1 = 70°C, \text{ liquid} \longrightarrow h_1 = 293.1\,\text{kJ/kg}, \ v_1 = 0.001023\,\text{m}^3/\text{kg}$$
$$T_2 = 25°C, \text{ saturated vapor} \longrightarrow h_2 = 2748\,\text{kJ/kg}$$

The mass flow rate of water is

$$\dot{m}_w = \frac{\dot{V}_1}{v_1} = \frac{(0.058/60)\ \text{m}^3/\text{s}}{0.001023\,\text{m}^3/\text{kg}} = 0.9451\,\text{kg/s}$$

The density of natural gas (assumed to be methane gas) at 25°C and 101 kPa is obtained from methane tables to be 0.6548 m^3/kg. Then, the rate of fuel consumed is

$$\dot{m}_{\text{fuel}} = \rho\dot{V} = (0.6548\,\text{kg/m}^3)\frac{500\,\text{m}^3}{2 \times 3600\,\text{s}} = 0.04547\,\text{kg/s}$$

The higher and lower heating values of natural gas are 55,500 and 50,000 kJ/kg, respectively. Then, the boiler efficiencies based on HHV and LHV are determined from

$$\eta = \frac{\dot{m}_w(h_2 - h_1)}{\dot{m}_{\text{fuel}}\text{HHV}} = \frac{(0.9451\,\text{kg/s})(2748 - 293.1)\ \text{kJ/kg}}{(0.04547\,\text{kg/s})(55,500\,\text{kJ/kg})} = 0.919 \text{ or } \textbf{91.9\%}$$

$$\eta = \frac{\dot{m}_w(h_2 - h_1)}{\dot{m}_{\text{fuel}}\text{LHV}} = \frac{(0.9451\,\text{kg/s})(2748 - 293.1)\ \text{kJ/kg}}{(0.04547\,\text{kg/s})(50,000\,\text{kJ/kg})} = 1.021 \text{ or } \textbf{102.1\%}$$

A 91.9% boiler efficiency indicates that 8.1% of the heating value of the fuel is lost mainly due to heat content of the exhaust gases and heat transfer from the boiler surfaces.

The boiler efficiency is greater than 100% when it is calculated based on the lower heating value. It is clear that in this boiler, the water in the combustion gases is condensed and thus some of the heat of vaporization is recovered. In order to avoid an efficiency value greater than 100%, the boiler efficiency should always be calculated based on the higher heating value.

4.1.7 Generator Efficiency and Overall Efficiency

A generator is a device that converts mechanical energy to electrical energy, and the effectiveness of a generator is characterized by the *generator efficiency*, which is the ratio of the electrical power output to the mechanical power input. The thermal efficiency of a power plant, which is of primary interest in thermodynamics, is usually defined as the ratio of the net shaft work output of the turbine to the heat input to the working fluid. The effects of other factors are incorporated by defining an *overall efficiency* for the power plant as the ratio of the net electrical power output to the rate of fuel energy input. That is,

$$\eta_{overall} = \eta_{combustion}\eta_{thermal}\eta_{generator} = \frac{\dot{W}_{net,electric}}{\dot{m}_{fuel}\text{HHV}} \tag{4.8}$$

4.1.8 Lighting Efficacy

We are all familiar with the conversion of electrical energy to light by incandescent lightbulbs, fluorescent tubes, and high-intensity discharge lamps. The efficiency for the conversion of electricity to light can be defined as the ratio of the energy converted to light to the electrical energy consumed. For example, common incandescent lightbulbs convert about 5% of the electrical energy they consume to light; the rest of the energy consumed is dissipated as heat, which adds to the cooling load of the air-conditioner in summer. However, it is more common to express the effectiveness of this conversion process by *lighting efficacy*, which is defined as the amount of light output in lumens per W of electricity consumed. The lighting efficacy of an ordinary incandescent lightbulb is between 6 and 20 lm/W and that of a compact fluorescent is between 40 and 90 lm/W. For white light sources the theoretical limit is about 300 lm/W [1].

4.2   Efficiencies of Mechanical and Electrical Devices

The transfer of mechanical energy is usually accomplished by a rotating shaft, and thus mechanical work is often referred to as *shaft work*. A pump or a fan receives shaft work (usually from an electric motor) and transfers it to the fluid as mechanical energy (fewer frictional losses). A turbine, on the other hand, converts the mechanical energy of a fluid to shaft work. In the absence of any irreversibilities such as friction, mechanical energy can be converted entirely from one mechanical form to another, and the *mechanical efficiency* of a device or process can be defined as

$$\eta_{mech} = \frac{\dot{E}_{mech,out}}{\dot{E}_{mech,in}} = 1 - \frac{\dot{E}_{mech,loss}}{\dot{E}_{mech,in}} \tag{4.9}$$

For example, the mechanical efficiency of a fan may be defined as

$$\eta_{mech,fan} = \frac{\Delta\dot{E}_{mech,fluid}}{\dot{W}_{shaft,in}} = \frac{\dot{m}V_2^2/2}{\dot{W}_{shaft,in}} \tag{4.10}$$

where $\dot{m}$ is the mass flow rate of air flowing through the casing of the fan, V_2 is the velocity of air at the fan exit, and $\dot{W}_{shaft,in}$ is the shaft power input to the fan. The overall efficiency of the fan may be defined as

$$\eta_{fan,\ overall} = \frac{\Delta\dot{E}_{mech,\ fluid}}{\dot{W}_{elect,\ in}} \tag{4.11}$$

where $\dot{W}_{elect,\ in}$ is the electrical power consumed by the fan. Similarly, the efficiency of a wind turbine may be expressed as

$$\eta_{wind\ turbine,\ overall} = \frac{\dot{W}_{elect,\ out}}{\Delta\dot{E}_{mech,\ fluid}} = \frac{\dot{W}_{elect,\ out}}{\dot{m}V_1^2/2} \tag{4.12}$$

where V_1 is the velocity of air at the turbine inlet and $\dot{W}_{elect,out}$ is the electrical power output from the turbine. Note that this efficiency is also equal to the exergy efficiency of a wind turbine because exergy expended is equal to the kinetic energy of the air at the turbine inlet.

A conversion efficiency of less than 100% indicates that conversion is less than perfect and some losses have occurred during conversion. A mechanical efficiency of 97% indicates that 3% of the mechanical energy input is converted to thermal energy as a result of frictional heating, and this will manifest itself as a slight rise in the temperature of the fluid.

In fluid systems, we are usually interested in increasing the pressure, velocity, and/or elevation of a fluid. This is done by supplying mechanical energy to the fluid by a pump, a fan, or a compressor (we refer to all of them as pumps). Or we are interested in the reverse process of extracting mechanical energy from a fluid by a turbine and producing mechanical power in the form of a rotating shaft that can drive a generator or any other rotary device. The degree of perfection of the conversion process between the mechanical work supplied or extracted and the mechanical energy of the fluid is expressed by the *pump efficiency* and *turbine efficiency*, defined as

$$\eta_{\text{pump}} = \frac{\Delta \dot{E}_{\text{mech,fluid}}}{\dot{W}_{\text{shaft,in}}} \tag{4.13}$$

$$\eta_{\text{turbine}} = \frac{\dot{W}_{\text{shaft,out}}}{\Delta \dot{E}_{\text{mech,fluid}}} \tag{4.14}$$

A pump or turbine efficiency of 100% indicates perfect conversion between the shaft work and the mechanical energy of the fluid, and this value can be approached (but never attained) as the frictional effects are minimized.

Electrical energy is commonly converted to *rotating mechanical energy* by electric motors to drive fans, compressors, robot arms, car starters, and so forth. The effectiveness of this conversion process is characterized by the *motor efficiency* η_{motor}, which is the ratio of the mechanical energy output of the motor to the electrical energy input. The full-load motor efficiencies range from about 35% for small motors to over 97% for large high-efficiency motors. The difference between the electrical energy consumed and the mechanical energy delivered is dissipated as waste heat.

The mechanical efficiency should not be confused with the *motor efficiency* and the *generator efficiency*, which are defined as

$$\eta_{\text{motor}} = \frac{\dot{W}_{\text{shaft,out}}}{\dot{W}_{\text{elect,in}}} \tag{4.15}$$

$$\eta_{\text{generator}} = \frac{\dot{W}_{\text{elect,out}}}{\dot{W}_{\text{shaft,in}}} \tag{4.16}$$

A pump is usually packaged together with its motor, and a turbine with its generator. Therefore, we are usually interested in the combined or overall efficiency of pump–motor and turbine–generator combinations, which are defined as

$$\eta_{\text{pump,overall}} = \eta_{\text{pump}}\eta_{\text{pump}} = \frac{\Delta \dot{E}_{\text{mech,fluid}}}{\dot{W}_{\text{elect,in}}} \tag{4.17}$$

$$\eta_{\text{turbine,overall}} = \eta_{\text{turbine}}\eta_{\text{generator}} = \frac{\dot{W}_{\text{elect,out}}}{\Delta \dot{E}_{\text{mech,fluid}}} \tag{4.18}$$

All the efficiencies just defined range between 0 and 100%. The lower limit of 0% corresponds to the conversion of the entire mechanical or electric energy input to thermal energy, and the device in this case functions as a resistance heater. The upper limit of 100% corresponds to the case of perfect conversion with no friction or other irreversibilities, and thus no conversion of mechanical or electric energy to thermal energy [1].

Example 4.2 Consider a hydroelectric power plant operating at an elevation of 55 m. The flow rate of water through the hydraulic turbine is 25,000 L/s. The electrical power output is measured as 1,500 kW. Determine the overall efficiency of the plant.

Solution The mechanical energy of the water is equal to its potential energy, which can be determined from the properties of water at the inlet and exit of the boiler and are obtained from steam tables to be

$$\Delta \dot{E}_{mech} = \dot{m}gz = \rho \dot{V}gz = (1\,kg/L)(2500\,L/s)(9.81\,m/s^2)(55\,m)\left(\frac{1\,kN}{1000kg \cdot m/s^2}\right)$$

$$= 1349\,kW$$

where the density of water is taken as 1 kg/L.

$$\eta_{turbine,overall} = \frac{\dot{W}_{elect,out}}{\Delta \dot{E}_{mech,fluid}} = \frac{1349\,kW}{1500\,kW} = 0.899 \text{ or } \mathbf{89.9\%}$$

4.3   Cryogenic Turbine Efficiencies

In conventional natural gas liquefaction plants, the high-pressure liquefied natural gas (LNG) is expanded in a throttling valve, also called a Joule–Thompson valve, where it undergoes a condensation process. The object of throttling is to decrease the LNG pressure to levels manageable for economic transportation and to allow the refrigeration cycle to be completed. A throttling process is essentially a constant enthalpy process and heat transfer to the fluid is negligible. From a thermodynamic point of view, a throttling valve can be replaced with a turbine. This way, the same pressure drop can be exploited by producing power. However, this replacement is often neither practical nor economical. A cryogenic turbine has been developed and tested by a private company for the replacement of the throttling valve in LNG liquefaction plants. An investigation of this cryogenic turbine revealed that replacing the throttling valve with the cryogenic turbine can save an LNG liquefaction plant about half a million dollars a year in electricity costs [15].

In this section, to establish a suitable model for the assessment of cryogenic turbine performance, the isentropic efficiency, the hydraulic efficiency, and the exergetic efficiency are studied and compared. LNG usually consists of about 99% methane and therefore the thermodynamic properties of pure methane were used. This section is based on Kanoglu [16].

Cryogenic turbines operate based on the hydraulic turbine principles. An hydraulic turbine extracts energy from a fluid, which possesses potential energy in the form of a high pressure head. The head represents the energy of a unit weight of the fluid.

Energy transfer occurs between the fluid in motion and a rotating shaft due to dynamic action, which results in changes in pressure and fluid momentum [17].

The cryogenic turbine is a radial inflow reaction turbine with an induction generator mounted on an integral shaft. The entire unit including the turbine and generator is submerged in the cryogenic liquid. In applications where a high power output is required, a radial turbine runner is more effective than a mixed flow or axial runner because the energy transfer between the fluid and the runner is enhanced by the reduction in radius along the fluid path through the runner. In addition, radial turbine runners have the advantage of lower runaway speeds when compared to mixed flow or axial runners. To expand high differential pressures, multistage turbines with multiple identical runners are utilized. Radial multistage turbines are both difficult to build and costly [18].

The isentropic efficiencies used to calculate the temperature changes across the cryogenic turbine are obtained from the following relation,

$$\eta_{hyd} = \frac{\dot{W}_{gen}}{\rho_{ave} g \dot{V} (\text{TDH})} \tag{4.19}$$

which is the defining relation for hydraulic efficiency. Theoretically isentropic and hydraulic efficiencies can be used interchangeably. The numerator in (4.19) is the actual power output from the generator and the denominator is the maximum possible power output. Density, volume flow rate, and total dynamic head are calculated from temperature and pressure measurements in the turbine. Below, the details of these calculations are given.

Densities at the turbine inlet and exit are calculated from the relations

$$\rho_1 = 1.692(T_{@\rho} - T_1) + \rho \tag{4.20}$$

$$\rho_2 = 1.692(T_{@\rho} - T_2) + \rho \tag{4.21}$$

where ρ is a known density at a given liquid temperature T. Theoretically, an infinite number of ρ and T combinations satisfy (4.20) and (4.21). The following combination is provided here to be used in these formulas.

$$\rho = 592.9 \, \text{kg/m}^3 \quad \text{and} \quad T = -49.76°\text{C}$$

This relation successfully accounts for the dependence of density with temperature. This empirical formula is obtained by correlating data. Density, as used in (4.19), is the average of these two densities. That is,

$$\rho_{ave} = \frac{\rho_1 + \rho_2}{2} \tag{4.22}$$

Volume flow rate is determined from

$$\dot{V} = C_V A_V \sqrt{\frac{\Delta P_V}{\rho_1}} \tag{4.23}$$

where the quantity C_V is the valve flow coefficient, tabulated in the manufacturers' brochures. The values of C_V in the literature increase nearly as the square of the size of the valve [19].

The total dynamic head is determined from

$$\text{TDH} = \frac{P_1 - P_2}{\rho_{ave} g} + \frac{\dot{V}^2}{2g}\left(\frac{1}{A_1^2} - \frac{1}{A_2^2}\right) + z_1 - z_2 \tag{4.24}$$

The heads due to velocity and elevation change are small, and the first term on the right-hand side of (4.24) dominates the calculation of total dynamic head.

Noticing the value of exergy analysis for all thermodynamic systems, the exergetic efficiencies of cryogenic turbines are also of importance. The exergetic efficiency of a turbine is defined as a measure of how well the stream exergy of the fluid is converted into actual turbine work output [11]. That is,

$$\varepsilon = \frac{w_{\text{out}}}{\psi_1 - \psi_2} \tag{4.25}$$

where work output from the turbine is

$$w_{\text{out}} = h_1 - h_2 \tag{4.26}$$

and the exergies of the fluid at the turbine inlet and exit are

$$\psi_1 = h_1 - h_0 - T_0(s_1 - s_0) \tag{4.27}$$

$$\psi_2 = h_2 - h_0 - T_0(s_2 - s_0) \tag{4.28}$$

Substituting (4.26) through (4.28) into (4.25), we find

$$\varepsilon = \frac{h_1 - h_2}{h_1 - h_2 - T_0(s_1 - s_2)} \tag{4.29}$$

Note that the arithmetic difference between the numerator and denominator in (4.29) is simply the specific exergy destruction in the turbine. The exergy destruction (irreversibility) due to heat loss from the turbine can be determined from

$$i_{\text{heat loss}} = q_{\text{out}}\left(1 - \frac{T_0}{T_1}\right) \tag{4.30}$$

Table 4.1 Operational data for testing facility

(a) Temperature change of LNG across the cryogenic turbine and the throttling valve (TV)

	T_1, °C	P_1, bar	P_2, bar	T_2, °C (Turbine)	T_2, °C (TV)	ΔT, °C (Turbine)	ΔT, °C (TV)
Test 1	−160.20	44.79	5.17	−159.54	−157.24	0.66	2.96
Test 2	−160.10	44.79	5.17	−159.38	−157.14	0.72	2.96
Test 3	−160.80	43.96	5.17	−159.98	−157.88	0.82	2.92
Test 4	−159.80	45.27	4.96	−158.84	−156.81	0.96	2.99
Test 5	−162.60	44.82	5.38	−161.39	−159.60	1.21	3.00

(b) Additional operational data for the cryogenic turbine

	Mass flow rate (kg/s)	Volume flow rate (m³/s)	Density (kg/m³)	Power (kW)	Isentropic eff. (η_s)	Exergy eff. (ε)
Test 1	129.22	0.2738	471.95	846.1	0.78	0.54
Test 2	130.47	0.2775	470.16	835.6	0.76	0.51
Test 3	137.73	0.2862	481.24	799.3	0.72	0.46
Test 4	117.95	0.2530	466.21	693.5	0.68	0.42
Test 5	110.80	0.2405	460.71	569.1	0.60	0.32

The heat loss from the cryogenic turbines is usually negligible. The rest of exergy destruction in the turbine is due to internal irreversibilities within the turbine, and can be determined from

$$i_{\text{internal}} = T_0 s_{\text{gen}} \qquad (4.31)$$

For the cryogenic turbine tested, the pressure and temperature at the turbine inlet (P_1 and T_1), the pressure at the turbine exit (P_2), and the isentropic efficiency of the turbine (η_s) are specified. The isentropic efficiency of an adiabatic turbine is given by

$$\eta_s = \frac{h_1 - h_2}{h_1 - h_{2s}} \qquad (4.32)$$

For cryogenic turbines, the isentropic efficiency, (4.32), and the hydraulic efficiency, (4.19), are essentially equal because the cryogenic liquid remains 100% liquid throughout the turbine and the enthalpy difference across the turbine in the isentropic case is essentially given by the denominator of (4.19). Note that the generator efficiency is included in the turbine efficiency so that the power output from the generator can be used in (4.19).

We now try to answer the question of whether isentropic efficiency or hydraulic efficiency is better suited for the determination of cryogenic turbine efficiency. The use of (4.32) is based on the determination of enthalpies from the measured pressure and temperature data at the turbine inlet and exit whereas (4.19) is based on the determination of generator power, density, volume flow rate, and total dynamic head. The determination of enthalpies requires the use of property relations.

A set of test operating data is provided by the testing facility in Tables 4.1 and 4.2. Table 4.1 gives temperature and other operational data. The data given in Table 4.2 include pressure and temperature measurements at the turbine inlet and exit and

Table 4.2 Isentropic, hydraulic, and exergetic efficiencies for the cryogenic turbine. The fluid is liquid propane

T_1, °C	P_1, bar	T_2, °C	P_2, bar	ΔT, °C	η_s	η_{hyd}	ε
Prototype test two-stage turbine							
Test 1	−50.12	10.98	−49.64	1.11	0.48	0.346	0.593	0.297
Test 2	−49.30	12.14	−48.81	1.10	0.49	0.402	0.518	0.348
Test 3	−51.10	8.03	−50.25	1.06	0.85	−0.570	0.474	−0.390
Test 4	−48.55	15.87	−47.81	1.10	0.74	0.331	0.464	0.279
Test 5	−50.44	16.81	−49.83	1.13	0.61	0.475	0.444	0.417
Test 6	−48.34	14.90	−47.42	1.08	0.92	0.119	0.279	0.094
Prototype test single stage turbine							
Test 1	−50.43	6.03	−50.74	1.10	−0.31	1.684	0.582	2.536
Test 2	−48.04	9.93	−47.70	1.15	0.34	0.469	0.571	0.420
Test 3	−48.00	9.75	−47.99	1.14	0.01	0.944	0.551	0.980
Test 4	−48.08	8.73	−47.95	1.11	0.13	0.742	0.433	0.722
Test 5	−51.57	8.75	−51.21	1.11	0.36	0.360	0.428	0.315
Test 6	−51.43	4.50	−51.03	1.06	0.40	−0.470	0.352	−0.345

hydraulic efficiencies determined using (4.19). The first six sets of test data are from a prototype two-stage turbine and the remaining six from a prototype single-stage turbine. This time, testing of the cryogenic turbine is done using liquid propane. However, the analysis should be equally applicable to the case of LNG because liquid propane exhibits similar cryogenic behavior through the turbine. The properties of propane are obtained from Klein [20]. Using the temperature and pressure measurements at the turbine inlet and exit, isentropic efficiencies are determined using (4.32). The results are obtained for 12 sets of test data and they are listed in Table 4.2 together with hydraulic efficiencies for comparison. Also listed in Table 4.2 are the temperature changes of propane across the turbine and the exergetic efficiencies. Environment temperature is taken to be 25°C in the calculation of exergetic efficiencies.

A quick observation of Table 4.2 shows that the isentropic efficiencies are not at all close to the hydraulic efficiencies. There seems to be no correlation between the two. Among 12 efficiencies calculated three are not thermodynamically possible because isentropic efficiency cannot be negative or greater than 1. The remaining nine are possible but grossly in error. This is also true for exergetic efficiencies. We now try to explain these results.

The temperature at the turbine exit is measured by a thermocouple whose accuracy is no better than ±0.33°C. The temperature at the turbine inlet is measured by a silicon diode thermometer with a much greater accuracy, ±0.04°C. Using a thermocouple at the turbine exit is due to practical and operational limitations. Both enthalpy and entropy are strong functions of temperature, and the temperature change of liquid propane across the turbine is extremely small, up to about 0.9°C. Consequently, the enthalpy and entropy changes of liquid propane across the turbine are small. With the temperatures measured, it is not possible to capture the exact enthalpy changes of the liquid across the turbine, thus causing erroneous results from the use of (4.32).

The uncertainties on pressure measurements at the turbine inlet and exit are estimated to be ±0.097% and ±0.53%, respectively. Compared with temperature, the uncertainties on pressure measurements have no significant effect on enthalpy and entropy calculations. This is because both the enthalpy and entropy are much more dependent on temperature than pressure. Note that the fluid undergoes a large pressure drop while experiencing very little temperature change.

A second factor that affects the isentropic efficiency calculation is the relations used in thermodynamic property evaluation. Such property relations are of limited accuracy and different references may use different property relations to determine the same fluid properties. The inaccuracies in property relations may have a significant effect on the isentropic efficiency calculation particularly when the temperature change, and thus the enthalpy change, is small as is the case in cryogenic turbines.

The argument given in the preceding paragraph for isentropic efficiency is equally valid for exergetic efficiency of cryogenic turbines. For actual thermodynamic systems, the second law (exergetic) efficiencies are greater than the first law efficiencies. Isentropic efficiency for turbines is also called the first law adiabatic efficiency. For steam turbines, the exergetic efficiencies are always greater than the isentropic efficiencies, the difference being small. It is therefore expected for the exergetic efficiencies to be greater than the isentropic efficiencies for the cryogenic turbines. The results listed in Table 4.2 show that this is not the case. This can be explained with the argument given in the preceding paragraph. The isentropic efficiency and exergetic efficiency results in Table 4.2 are obtained using the same reference [20] for property evaluation and these efficiencies should not be taken literally.

To provide yet another comparison between the isentropic and exergetic efficiencies of cryogenic turbines, exergetic efficiencies for the five sets of test data provided in Table 4.1 are calculated. The results are listed in Table 4.1b and depicted in Fig. 4.1. The exergetic efficiencies range between 0.54 and 0.32 whereas the isentropic efficiencies range between 0.78 and 0.60. It appears that the exergetic efficiencies are on average 0.26 smaller than the isentropic efficiencies. This is again in contrast to the steam turbines. For the same reasons explained, these results should not be taken literally. However, without drawing any conclusion, we notice the trend that isentropic efficiencies are greater than the exergetic efficiencies.

We have a good level of confidence in the hydraulic efficiency inasmuch as its calculation is based on the determination of generator power, density, volume flow rate, and total dynamic head and they all can be determined with reasonable accuracies. To check the validity of turbine hydraulic efficiencies, effects of uncertainties in the measurements of temperature, pressure, and generator power on the turbine hydraulic efficiency are studied based on test operating data, and the uncertainty in turbine hydraulic efficiency is estimated to be ±0.20%. This uncertainty in the turbine efficiency appears to be reasonable and acceptable in the assessment of turbine performance. The details of the uncertainty analysis of the cryogenic turbines as well as certain thermodynamics aspects of cryogenic turbines can be found in [21].

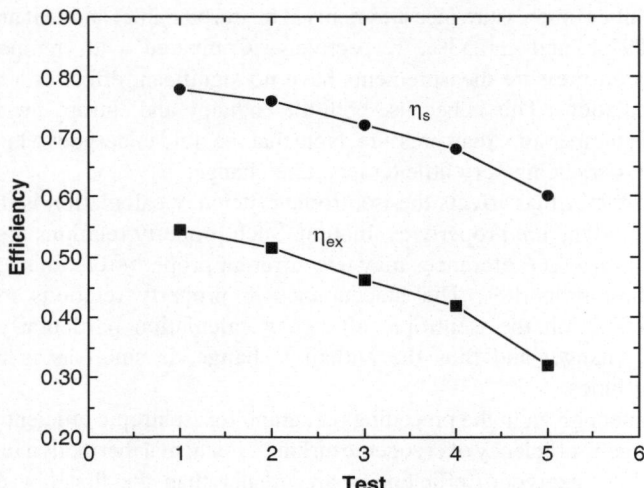

Fig. 4.1 Comparison of isentropic and exergetic efficiencies of cryogenic turbine for five sets of test data. The fluid is LNG and the values are from Table 4.1(b)

It should be clear from the results that the cryogenic turbine efficiency should be determined using the hydraulic efficiency and not the isentropic efficiency. However, the isentropic efficiency (and/or the exergetic efficiency) is a good way to assess performance of the gas and steam turbines because the difference in the fluid temperature across the turbine is high, and so are the enthalpy differences. As a result, effects of the temperature measurement uncertainties and the inaccuracies involved in property relations on the isentropic efficiency are usually insignificant. For fluids other than liquids, the enthalpy difference in the isentropic case cannot be expressed by the expression given in the denominator of (4.19). Therefore, for the gas and steam turbines we have to rely on isentropic efficiency (and/or exergetic efficiency) to assess the turbine performance.

Chapter 5
Efficiencies of Power Plants

5.1 Introduction

To assist in improving the efficiencies of power plants, their thermodynamic characteristics and performance are usually investigated. Power plants are normally examined using energy analysis but, as pointed out previously, a better understanding is attained when a more complete thermodynamic view is taken, which uses the second law of thermodynamics in conjunction with energy analysis via exergy methods.

Although exergy analysis can be generally applied to energy and other systems, it appears to be a more powerful tool than energy analysis for power cycles because it helps determine the true magnitudes of losses and their causes and locations, and improve the overall system and its components. In this chapter, we provide an overview of various energy- and exergy-based efficiencies used in the analysis of power cycles, including vapor and gas power, cogeneration and geothermal power plants. Differences in design aspects are considered. The various approaches that can be used in defining efficiencies are identified and their implications discussed. Numerical examples are provided to illustrate the use of the different efficiencies, and the results include combined energy and exergy diagrams.

Note that the emphasis in this chapter is to describe various energy- and exergy-based efficiencies used in power plants and discuss the implications associated with each definition. Therefore, simple cycles are selected to keep the complexity of the plants at a minimum level for gas and vapor cycles the better to facilitate understanding of the efficiencies, which can be very useful for improved energy management in power plants. One can easily adapt the efficiencies discussed here to more complex power systems. Some efficiency definitions for gas cycles found in many thermodynamics textbooks are repeated so that the coverage in this chapter is comprehensive and can serve as a convenient and practical tool for students, engineers, and researchers. It is shown that a better understanding of energy and exergy efficiencies and their successful use can help improve energy management in power plants [6].

M. Kanoğlu et al., *Efficiency Evaluation of Energy Systems*,
SpringerBriefs in Energy, DOI 10.1007/978-1-4614-2242-6_5,
© Mehmet Kanoğlu, Yunus A. Çengel, İbrahim Dinçer 2012

5.2 Efficiencies of Vapor Power Cycles

The thermal efficiency, also referred to as the energy efficiency or the first law efficiency, of a power cycle is defined as

$$\eta_{th-1} = \frac{w_{net,\,out}}{q_{in}} = 1 - \frac{q_{out}}{q_{in}} \tag{5.1}$$

where $w_{net,out}$ is the specific net work output, q_{out} is the specific heat rejected from the cycle, and q_{in} is the specific heat input to the cycle, which is usually taken to be the specific heat input to the steam in the boiler of a steam power plant. That is,

$$q_{in} = h_3 - h_2 \tag{5.2}$$

where h denotes specific enthalpy and the subscripts refer to state points in Fig. 5.1. This simple approach neglects the losses occurring in the furnace–boiler system due to the energy lost with hot exhaust gases, incomplete combustion, and so on. To incorporate these losses, one can express the thermal efficiency of the cycle by a second approach as

$$\eta_{th-2} = \frac{\dot{W}_{net,\,out}}{\dot{m}_{fuel}q_{HV}} \tag{5.3}$$

where $\dot{m}_{fuel}$ is the mass flow rate of fuel and q_{HV} is the heating value of the fuel, which can be chosen as the higher or lower heating value. For furnace–boiler systems where the water in the exhaust gases is not expected to condense, as in internal combustion engines, it is customary to use the lower heating value [22].

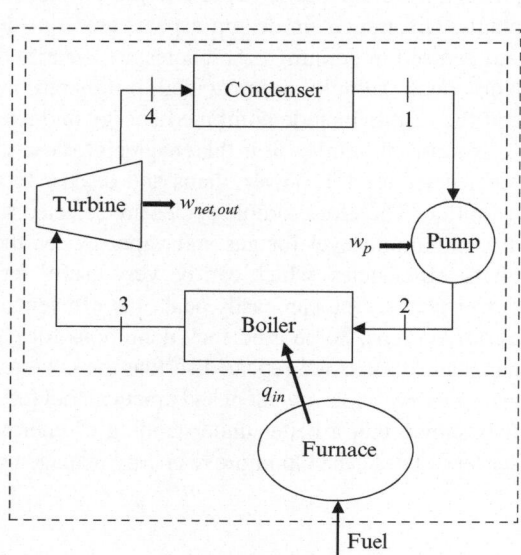

Fig. 5.1 Simple steam power plant

Some tend to use lower heating values to make a device appear more efficient. This is frequently done in manufacturer descriptions of commercial boilers. Often a claimed efficiency exceeds 100%, as discussed in Chap. 4. This is due to recovering some of the heat of condensation of steam in the exhaust gases while still defining boiler efficiency based on the lower heating value. This is misleading and a thermodynamically improper use of efficiency. If there is any possibility of recovering some of the energy of condensing steam in exhaust gases, the efficiency should be based on the higher heating value.

The second law efficiency, also referred to as exergy efficiency, of a power-producing cycle is defined as

$$\varepsilon = \frac{w_{\text{net, out}}}{x_{\text{in}}} = 1 - \frac{x_{\text{dest}}}{x_{\text{in}}} \tag{5.4}$$

where x_{in} is the specific exergy input to the cycle and x_{dest} is the specific total exergy destruction in the cycle. One can express the exergy input to the cycle as the exergy increase of the working fluid in the boiler of a steam power plant (Fig. 5.1) as

$$x_{\text{in}} = h_3 - h_2 - T_0(s_3 - s_2) \tag{5.5}$$

where T_0 is the dead-state or environment temperature and s is the specific entropy. Substituting (5.5) into (5.4),

$$\varepsilon_1 = \frac{w_{\text{net, out}}}{h_3 - h_2 - T_0(s_3 - s_2)} \tag{5.6}$$

In this definition, the irreversibilities during energy transfer from the furnace to the steam in the boiler are not accounted for. Alternatively, the exergy input to the cycle may be defined as the exergy input to the boiler accompanying the heat transfer. The exergy efficiency in this case becomes

$$\varepsilon_2 = \frac{w_{\text{net, out}}}{q_{\text{in}}\left(1 - \dfrac{T_0}{T_s}\right)} \tag{5.7}$$

where T_s is the source temperature, which is the temperature of the heat source (i.e., furnace), and q_{in} is given by (5.2). This efficiency definition incorporates the irreversibility during heat transfer to the steam in the boiler. We may also incorporate in the efficiency definition the exergy destruction associated with fuel combustion and the exergy lost with exhaust gases from the furnace. In this third approach, the exergy efficiency can be expressed as

$$\varepsilon_3 = \frac{\dot{W}_{\text{net, out}}}{\dot{m}_{\text{fuel}}x_{\text{fuel}}} \tag{5.8}$$

where x_{fuel} is the specific exergy of the fuel. The exergy of a fuel may be obtained by writing the complete combustion reaction of the fuel and calculating the reversible work by assuming all products are at the state of the surroundings. Then the exergy of fuel is equivalent to the calculated reversible work. For fuels whose combustion reaction involves water in the products, the exergy of the fuel is different depending on the phase of the water (vapor or liquid). The exergies of various fuels listed in [10] are based on the vapor phase of water in combustion gases.

Different efficiency definitions are possible if one selects different system boundaries. Clearly defining the system boundary allows the efficiency to be defined unambiguously. For example, the exergy efficiencies in (5.7) and (5.8) correspond to systems whose boundaries are given by the inner and outer dashed lines, respectively, in Fig. 5.1.

Example 5.1 A numerical example is used to illustrate and contrast the various efficiencies defined in this section. We consider a simple steam power plant with a net power output of 10 MW and boiler and condenser pressures of 10,000 and 10 kPa, respectively (Fig. 5.1). We assume a turbine inlet temperature of 500°C and isentropic efficiencies of 85% for both the turbine and the pump. In addition, we assume that the furnace–boiler system has an efficiency of 75%. That is, 75% of the lower heating value of the fuel is transferred to the steam flowing through the boiler and the remaining 25% is lost, mostly with the hot exhaust gases passing through the chimney. The source and sink temperatures in (5.7) are taken as 1,300 and 298 K, respectively. We consider methane as the fuel with a lower heating value of 50,050 kJ/kg and a chemical exergy of 51,840 kJ/kg [10].

For the given values and assumptions, an analysis of this cycle yields

$$w_{net,\,out} = 1,081\,\text{kJ/kg}, \; q_{in} = 3,172\,\text{kJ/kg}, \; x_{in-1} = 1,400\,\text{kJ/kg}, \; x_{in-2} \\ = 2,444\,\text{kJ/kg}$$

as well as the following efficiency values:

$$\eta_{th-1} = 34.1\% \; , \; \eta_{th-2} = 25.6\%, \; \varepsilon_1 = 77.2\%, \; \varepsilon_2 = 44.2\%, \; \varepsilon_3 = 24.7\%$$

When the energy and exergy losses in the furnace–boiler system are not considered, the thermal efficiency is 34.1% whereas the corresponding exergy efficiency is much higher (77.2%). However, when the losses in the furnace–boiler are considered, the exergy efficiency (24.7%) is lower than the thermal efficiency (25.6%). When teaching undergraduate thermodynamics, it is normally stated that the exergy efficiency is greater than the thermal efficiency for heat engines, referring to the first approach here. This point is made by emphasizing that thermal efficiency is the fraction of heat input that is converted to work whereas exergy efficiency is the fraction of the work potential of the heat (this work potential, i.e., exergy, is smaller than heat) that is converted to work. However, when one considers the effect of furnace–boiler losses, and uses the chemical exergy of the fuel in the exergy efficiency and the heating value of the fuel in the thermal efficiency, the exergy efficiency becomes smaller than the

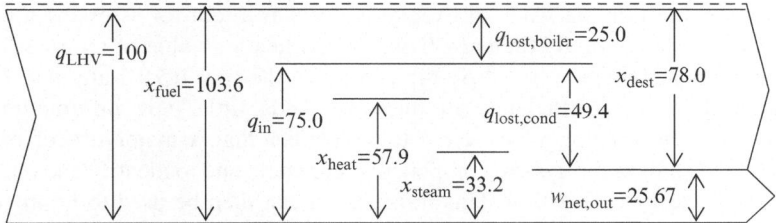

Fig. 5.2 Combined energy and exergy diagram for the steam power plant considered

thermal efficiency. In thermodynamics, it is often misleading to make generalized statements as they may not always apply. For example, can we state that the exergy efficiency, based on the third approach in (5.8) (ε_3), is always lower than the thermal efficiency as defined by the second approach in (5.3) ($\eta_{th\text{-}2}$)? The answer will be yes only if the chemical exergy of the fuel is always greater than its heating value. According to data in [10], this is the case for methane but not for hydrogen (q_{LHV} = 119,950 kJ/kg, x_{fuel} = 117,120 kJ/kg).

For a reversible heat engine cycle operating between a source at T_s and a sink at T_0, the thermal efficiency is given by

$$\eta_{th,\,rev} = 1 - \frac{T_0}{T_s} \tag{5.9}$$

The ratio of the actual thermal efficiency to the thermal efficiency of a reversible heat engine operating between the same temperature limits gives a type of exergy efficiency of the heat engine. For a furnace temperature of $T_s = 1,300$ K and an environment temperature of $T_0 = 298$ K, the reversible thermal efficiency found with (5.9) is 77.1%. Dividing the actual thermal efficiency of 34.1% by this efficiency (0.341/0.771) gives 44.2%. Note that this is the same as the exergy efficiency obtained using (5.7).

The results of the numerical example considered in this section are shown in a combined energy and exergy diagram in Fig. 5.2. In many studies with energy and exergy analyses of power cycles, energy and exergy flow diagrams are given separately. The combined flow diagram approach used here appears to be useful in conveying energy and exergy results of the cycle in a scaled, compact, and comprehensive manner. The heating value of the fuel is normalized to 100 units of energy and other values are normalized accordingly. The thermal and exergy efficiencies discussed in this section can easily be obtained using the values in this diagram by taking the ratios of various terms. The total exergy destruction in this power plant is 78 kJ for a total exergy input of 103.6 kJ. The exergy destruction in the cycle based on an exergy input of 33.2 kJ is only 7.6 kJ (33.2–25.6), which is only 9.7% of total exergy destruction. That is, the exergy destructions in the furnace–boiler system account for the remaining 90.3% of the total exergy destruction. This significant exergy destruction is not considered in an exergy efficiency definition neglecting the destructions in the furnace–boiler system [see (5.4)].

One may question the value of exergy analysis as a tool for assessing a power plant because the thermal efficiency based on the heating value of the fuel [(5.3)] and the exergy efficiency based on the exergy of the fuel [(5.8)] are very close. Although the exergy efficiency in this case adds little new information for addressing cycle efficiency, we have to remember that a major use of exergy analysis is to analyze the system components separately and to identify and quantify the sites of exergy destruction. This information can then be used to improve the performance of the system by trying to minimize the exergy destructions in a prioritized manner. Note that the exergy efficiency defined in (5.6) addresses the fact that only a fraction of the heat from combustion that is transferred to the steam in the boiler is available for work, and the exergy efficiency compares the actual work output to this available work (i.e., exergy). The exergy efficiencies in these cases become greater than the corresponding thermal efficiencies, providing more realistic measures of system performance compared to the corresponding thermal efficiencies. For a more comprehensive thermodynamic analysis of a power cycle, the various energy- and exergy-based efficiencies are best considered.

5.3   Efficiencies of Gas Power Plants

The schematic of an open-cycle gas turbine power plant is given in Fig. 5.3. The thermal efficiency of this plant may be expressed as

$$\eta_{th} = \frac{\dot{W}_{net,\,out}}{\dot{m}_{fuel} q_{HV}} \qquad (5.10)$$

The exergy efficiency of this gas-turbine engine is

$$\varepsilon = \frac{\dot{W}_{net,\,out}}{\dot{m}_{fuel} x_{fuel}} \qquad (5.11)$$

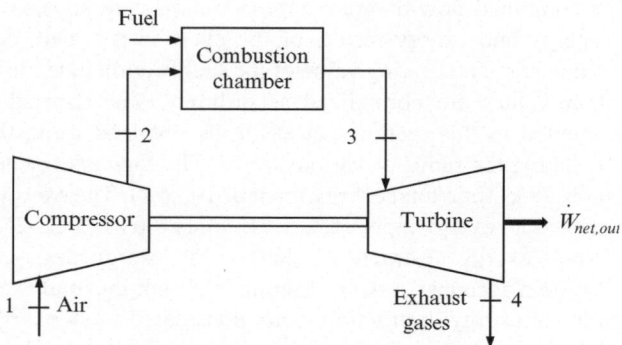

Fig. 5.3 An open-cycle gas-turbine engine

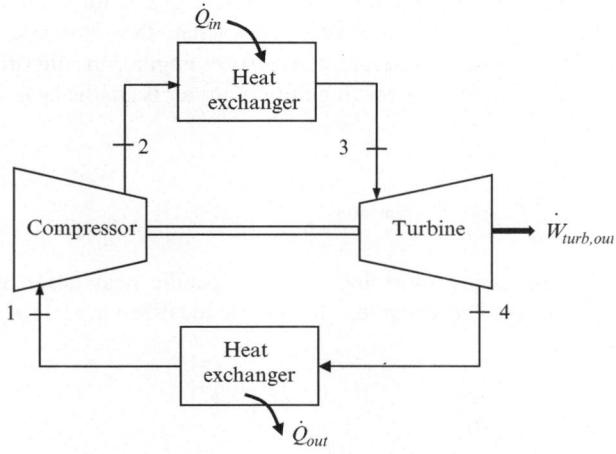

Fig. 5.4 A closed-cycle gas-turbine engine

This engine is sometimes modeled by a closed-cycle gas-turbine engine as shown in Fig. 5.4. The working fluid is assumed to be air and the combustion process is replaced by a heat addition process. In this cycle, the energy efficiency may be written as

$$\eta_{th} = \frac{\dot{W}_{net,\,out}}{\dot{m}_{fuel}q_{HV}} = \frac{\dot{W}_{net,\,out}}{\dot{m}_{air}q_{in}} = \frac{\dot{W}_{turb,\,out} - \dot{W}_{comp,\,in}}{\dot{m}_{air}(h_3 - h_2)} \tag{5.12}$$

Note that the heat added to the cycle is equal to the heat resulting from the combustion process. Equation 5.12 is equivalent to (5.10). The exergy efficiency may be written by different approaches as

$$\varepsilon = \frac{\dot{W}_{net,\,out}}{\dot{Q}_{in}\left(1 - \frac{T_0}{T_s}\right)} \tag{5.13}$$

$$\varepsilon = \frac{\dot{W}_{net,\,out}}{\dot{X}_3 - \dot{X}_2} = \frac{\dot{W}_{net,\,out}}{\dot{m}_{air}[h_3 - h_2 - T_0(s_3 - s_2)]} \tag{5.14}$$

Here, (5.11, 5.13), and (5.14) give different results for the exergetic efficiency values. Equation 5.13 does not account for the exergy destruction during the combustion process whereas (5.14) does not account for the exergy destructions during combustion and during the heat transfer to the working fluid in the cycle. The efficiency will be highest in (5.14) and lowest in (5.11).

Energy and exergy efficiencies of an actual internal combustion engine can be expressed using (5.10) and (5.11). For the idealized models of internal combustion engines (Otto, Diesel, and Dual cycles), modified versions of (5.12) through (5.14) may easily be obtained using the same principles.

Simplified thermal efficiency relations of idealized cycles for internal combustion engines and gas-turbine cycles are available. When the Otto cycle is used to represent the operation of an internal combustion engine, the thermal efficiency under air-standard assumptions (working fluid is air; air is an ideal gas with constant specific heats) is

$$\eta_{th,\ Otto} = 1 - \frac{1}{r^{k-1}} \tag{5.15}$$

where r is the compression ratio and k is the specific heat ratio. Similarly, the thermal efficiency of the Diesel cycle, which is the idealized model for compression ignition engines, is

$$\eta_{th,\ Diesel} = 1 - \frac{1}{r^{k-1}} \left[\frac{r_c^k - 1}{k(r_c - 1)} \right] \tag{5.16}$$

where r_c is the cutoff ratio, defined as the ratio of cylinder volumes after and before the combustion process. The efficiency relation for the Dual cycle is

$$\eta_{th,\ Dual} = 1 - \frac{1}{r^{k-1}} \left[\frac{r_p r_c^k - 1}{k r_p (r_c - 1) + r_p - 1} \right] \tag{5.17}$$

where r_p is the ratio of pressures after and before the constant-volume heat addition process. The thermal efficiency of the simple Brayton cycle, which is the idealized model for gas-turbine engines, is expressed using the air-standard assumption as

$$\eta_{th,\ Brayton} = 1 - \frac{1}{r_p^{(k-1)/k}} \tag{5.18}$$

where r_p is the ratio of maximum and minimum pressures in the cycle. For the idealized regenerative Brayton cycle, the efficiency relation is

$$\eta_{th,\ Brayton,\ regen} = 1 - \left(\frac{T_1}{T_3} \right) r_p^{(k-1)/k} \tag{5.19}$$

where T_1 and T_3 are the temperatures at the inlets of the compressor and the turbine, respectively.

The operational description of these idealized cycles may be found in most thermodynamics textbooks [1, 23]. Equations 5.15 through 5.19 are only applicable to the idealized cycles considered, and they should not be used to determine the thermal efficiencies of actual internal combustion engines or gas-turbine cycles. Equations 5.15 through 5.19 are useful in that they illustrate the effects of some key design parameters such as compression ratio, cutoff ratio, and pressure ratio on cycle efficiency.

5.4 Efficiencies of Cogeneration Plants

Cogeneration refers to the simultaneous generation of more than one form of energy product. Performance assessment of various cogeneration plants are given in Kanoglu and Dincer [24]. For a cogeneration plant producing electric power $\dot{W}_{\text{net,out}}$ and process heating $\dot{Q}_{\text{process}}$, a first-law-based efficiency is defined as the ratio of useful energy output to energy input:

$$\eta_{\text{cogen}} = \frac{\dot{W}_{\text{net, out}} + \dot{Q}_{\text{process}}}{\dot{Q}_{\text{in}}} = 1 - \frac{\dot{Q}_{\text{loss}}}{\dot{Q}_{\text{in}}} \qquad (5.20)$$

where $\dot{Q}_{\text{process}}$ is the output rate of process heat and $\dot{Q}_{loss}$ is the heat lost in the condenser. This relation is referred to as the utilization efficiency to differentiate it from the thermal efficiency which is used for a power plant where the single output is power. Students are consistently taught not to compare apples and oranges, which usually refers to two commodities that are different. Work and heat have the same units but are fundamentally difficult to add because they are different, with work being a more valuable commodity than heat.

We can overcome this situation by defining the efficiency of a cogeneration plant based on exergy, as the ratio of total exergy output to exergy input:

$$\varepsilon_{\text{cogen}} = \frac{\dot{X}_{\text{out}}}{\dot{X}_{\text{in}}} = \frac{\dot{W}_{\text{net, out}} + \dot{X}_{\text{process}}}{\dot{X}_{\text{in}}} = 1 - \frac{\dot{X}_{\text{dest}}}{\dot{X}_{\text{in}}} \qquad (5.21)$$

where $\dot{X}_{\text{process}}$ is the exergy transfer rate associated with the transfer of process heat, expressible as

$$\dot{X}_{\text{process}} = \int \delta\dot{Q}_{\text{process}}\left(1 - \frac{T_0}{T}\right) \qquad (5.22)$$

where T is the instantaneous source temperature from which the process heat is transferred. This relation is of little practical value unless the functional relationship between the process heat rate $\dot{Q}_{\text{process}}$ and temperature T is known. In many cases, the process heat is utilized by the transfer of heat from a working fluid exiting the heat producing device (e.g., a turbine or an internal combustion engine) to a secondary fluid in a heat exchanger (Fig. 5.5). One can express the exergy rate of process heat as the exergy decrease of the hot fluid in the heat exchanger as

$$\dot{X}_{\text{process}-1} = -\Delta\dot{X}_{\text{hot}} = \dot{m}_{\text{hot}}[h_1 - h_2 - T_0(s_1 - s_2)]_{\text{hot}} \qquad (5.23)$$

or by the increase of the exergy of the cold fluid in the heat exchanger

$$\dot{X}_{\text{process}-2} = \Delta\dot{X}_{\text{cold}} = \dot{m}_{\text{cold}}[h_4 - h_3 - T_0(s_4 - s_3)]_{\text{cold}} \qquad (5.24)$$

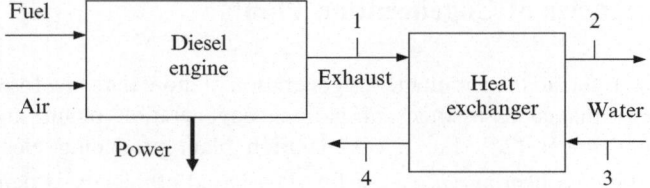

Fig. 5.5 Cogeneration plant with a diesel engine and a heat exchanger for steam production

where the subscripts refer to state points in Fig. 5.5. The difference between these two exergies is the exergy destruction in the heat exchanger. Then, from (5.21), the exergy efficiencies based on these two approaches become

$$\varepsilon_{cogen-1} = \frac{\dot{W}_{net,\,out} + \dot{m}_{hot}[h_1 - h_2 - T_0(s_1 - s_2)]_{hot}}{\dot{X}_{in}} \tag{5.25}$$

and

$$\varepsilon_{cogen-2} = \frac{\dot{W}_{net,\,out} + \dot{m}_{cold}[h_4 - h_3 - T_0(s_4 - s_3)]_{cold}}{\dot{X}_{in}} \tag{5.26}$$

The exergy input in these relations can be expressed differently using various inputs as in the denominators of (5.6) through (5.8), yielding different exergy efficiencies.

Example 5.2 To illustrate the use of these efficiencies, we consider a diesel engine-based cogeneration plant, The outputs are electrical power and process heat, which is transferred from the hot exhaust gases to water to produce steam in a heat exchanger (Fig. 5.5). Some of the data used in this example are from an actual diesel engine power plant [25]. The net power output from the plant is 18,900 kW when the rate of fuel consumption rate is 1.03 kg/s and the air–fuel ratio is 40.4. This corresponds to an exhaust flow rate of 41.6 kg/s. The plant uses heavy diesel fuel with a lower heating value of 39,300 kJ/kg. The exhaust gases enter the process heating unit (i.e., heat exchanger) at 383°C and experience a temperature drop of 175°C whereas compressed liquid water enters at 15°C and exits as saturated vapor at 200°C. Applications of (5.20) through (5.26) produce the following results.

$$\dot{Q}_{process} = 7784\,kW, \quad \dot{X}_{in} = 43,110\,kW, \quad \dot{X}_{process-1} = 3678\,kW$$

$$\dot{X}_{process-2} = 2509\,kW, \quad \eta_{cogen} = 65.9\%, \quad \varepsilon_{cogen-1} = 52.4\%, \quad \varepsilon_{cogen-2} = 49.7\%$$

The exergy of heavy diesel fuel with an unknown composition is taken as 1.065 times the lower heating value of the fuel following the approach by Brzustowski and Brena [26]. Properties of air with variable specific heats are used for exhaust gases.

The difference between the energy and exergy efficiencies in this cogeneration plant appears to be much greater than the difference for a power plant, when the

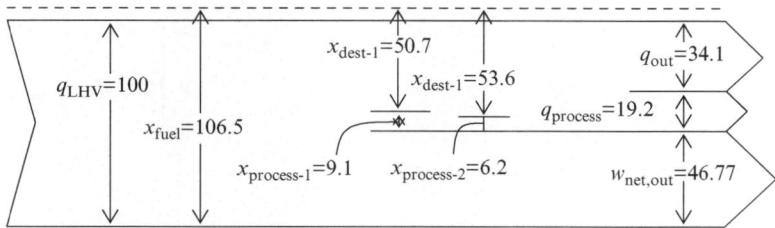

Fig. 5.6 Combined energy and exergy diagram for the cogeneration plant considered

energy and exergy efficiencies are respectively defined based on the energy and exergy of the fuel, as discussed in the previous section. The difference is attributable to one of the product outputs being process heat. The different approaches used in (5.18) and (5.19) to define the exergy of the process heat result in a small exergy efficiency difference of only $52.4 - 49.7 = 2.7\%$. The greater the average temperature difference between the hot and cold fluids in the process heater, the greater is the exergy destruction and the greater is the difference between the two definitions of exergy efficiencies in (5.25) and (5.26), respectively.

The results are presented in a combined energy and exergy diagram in Fig. 5.6. The heating value of the fuel (i.e., heat input) is normalized to 100 units of energy and other values are modified accordingly. The energy and exergy efficiencies discussed in this section can be found using this diagram. The total exergy destruction in this cogeneration plant is 50.7 kJ based on the first approach and 53.6 kJ based on the second approach, for a total exergy input of 106.5 kJ. The difference between these exergy destructions is the exergy destruction in the process heater, which is 5.7% of total exergy destruction or 2.7% of exergy input.

5.4.1   Steam-Turbine-Based Cogeneration Plant

Referring to Fig. 5.7 for the states, the energy efficiency is expressible as

$$\eta_{cogen} = \frac{\dot{W}_{net} + \dot{m}_{water}(h_{10} - h_9)}{\dot{m}_{fuel}q_{LHV}} \tag{5.27}$$

where $\dot{m}_{fuel}$ is the mass flow rate of fuel and q_{LHV} is the lower heating value of the fuel. Higher heating value can also be used in this equation. Using a higher heating value would correspond to a lower energy efficiency compared to using a lower heating value.

The value of the mass flow rate of the steam extracted from the turbine at state 5 (Fig. 5.7) affects the energy efficiency of the cogeneration plant. The greater the amount of mass extracted is, the greater the amount of heating and the smaller the

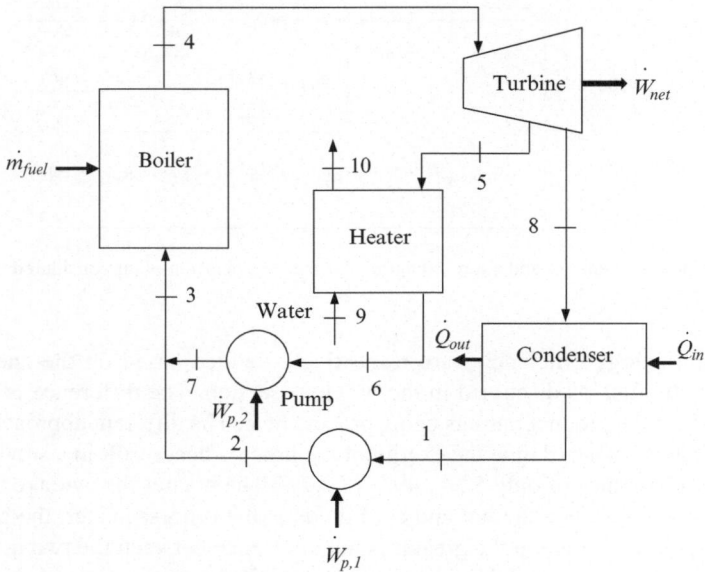

Fig. 5.7 A steam-turbine cogeneration plant

power output. The optimum value should be selected depending on the heat and power demands.

The exergy input to a steam cogeneration plant can be expressed as

$$\dot{X}_{in} = \dot{m}_{fuel} x_{fuel} \tag{5.28}$$

where x_{fuel} is the specific exergy of the fuel. The exergy of a fuel may be obtained by writing the complete combustion reaction of the fuel and calculating the reversible work obtainable assuming all products are at the state of surroundings. The exergy of the fuel is equal to the reversible work. For fuels that yield water as a combustion product, the exergy of the fuel differs depending on the water's phase (vapor or liquid). Szargut et al. [10] list the exergies of various fuels based on the vapor phase of water in combustion gases.

Referring to the states as given in Fig. 5.7, we obtain the exergy efficiency as

$$\varepsilon_{cogen} = \frac{\dot{W}_{net} + \dot{m}_{water}[h_{10} - h_9 - T_0(s_{10} - s_9)]}{\dot{m}_{fuel} x_{fuel}} \tag{5.29}$$

As an illustrative example, we consider a cogeneration steam power plant with a net power output of 10 MW and boiler and condenser pressures of 10,000 and 10 kPa, respectively (Fig. 5.7). The turbine inlet temperature is 500°C and isentropic efficiencies for both the turbine and the pump are assumed to be 85%. Steam is extracted from the turbine at 2 MPa pressure and used to obtain hot water for the radiator from the heater. The steam exits the heater at the same pressure as a saturated liquid. The liquid water, heated to 90°C, is used to heat buildings and returns to the cogeneration plant at 50°C as a common practice.

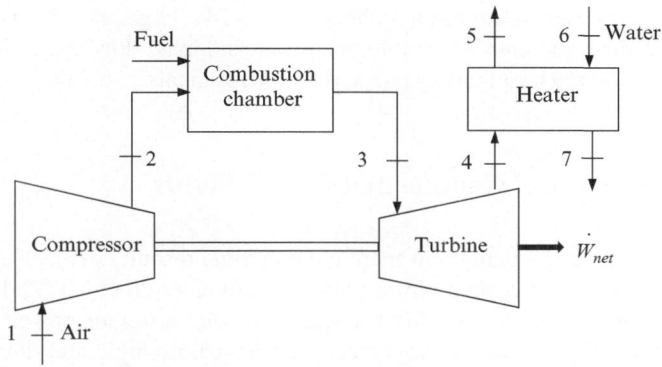

Fig. 5.8 A gas-turbine cogeneration plant

The energy efficiency of the furnace–boiler system is taken to be 90%. That is, 90% of the lower heating value of the fuel is transferred to the steam generated in the boiler and the remaining 10% is lost, mostly with the hot exhaust gases passing through the chimney. The dead-state temperature is taken to be 25°C. We consider methane as the fuel with a lower heating value of 50,050 kJ/kg and a chemical exergy of 51,840 kJ/kg [10].

For a net power output of 10 MW, the mass flow rate through the boiler is 16.17 kg/s and the extracted steam mass flow from the turbine for heating water is 6.78 kg/s. Under these operating conditions, the mass flow rate of liquid water is 80.5 kg/s and the corresponding heat transfer in the heater is 13.5 MW.

5.4.2   Gas-Turbine-Based Cogeneration Plant

The energy efficiency of a gas-turbine cogeneration plant can be expressed as

$$\eta_{cogen} = \frac{\dot{W}_{net} + \dot{m}_{water}(h_7 - h_6)}{\dot{m}_{fuel}q_{LHV}} \tag{5.30}$$

where $\dot{m}_{water}$ is the mass flow rate of water and the state numbers are shown in Fig. 5.8. The exergy efficiency is given by

$$\varepsilon_{cogen} = \frac{\dot{W}_{net} + \dot{m}_{water}[h_7 - h_6 - T_0(s_7 - s_6)]}{\dot{m}_{fuel}x_{fuel}} \tag{5.31}$$

As an illustrative example, we consider a cogeneration gas-turbine power plant with a net power output of 10 MW and maximum and minimum pressures of 1,200 and 100 kPa in the system, respectively (Fig. 5.8). The fuel is methane, the turbine inlet temperature is 700°C and the isentropic efficiencies of both the turbine and compressor are 85%. Under these operating conditions, exhaust gases leave the turbine at 303°C and the mass flow rate steam through the turbine is 141.9 kg/s. For the water heated in the heater, the same inlet and exit temperatures and mass flow rate as in Sect. 5.3 are considered, so the heat transfer rate in the heater remains 13.5 MW.

5.5   Efficiencies of Geothermal Power Plants

The technology for producing power from geothermal resources is well established and there are many geothermal power plants operating worldwide [27]. Depending on the state of the geothermal fluid in the reservoir, different power-producing cycles may be used including direct steam, flash-steam (single and double-flash), binary and combined flash–binary cycles. In general, the thermal efficiency of a geothermal power plant may be expressed as

$$\eta_{\text{th}-1} = \frac{\dot{W}_{\text{net, out}}}{\dot{E}_{\text{in}}} \tag{5.32}$$

where $\dot{E}_{\text{in}}$ is the energy input rate to the power plant, which may be expressed as the specific enthalpy of the geothermal water with respect to the environment state multiplied by the mass flow rate of geothermal water $\dot{m}_{\text{geo}}$. That is,

$$\eta_{\text{th}-1} = \frac{\dot{W}_{\text{net, out}}}{\dot{m}_{\text{geo}}(h_{\text{geo}} - h_0)} \tag{5.33}$$

The state of the geothermal water may be taken as that in the reservoir or at the well head. Those who use the reservoir state argue that a realistic and more meaningful comparison between geothermal power plants needs to account for methods of harvesting the geothermal fluid. However, those who use the well-head-state argue that taking the reservoir as the input is not appropriate for geothermal power plants because conventional power plants are evaluated on the basis of the energy of the fuel burned at the plant site [28–30]. In (5.33), the energy input to the power plant represents the maximum heat the geothermal water can deliver, which occurs when the geothermal water is cooled to the temperature of the environment.

The simplest geothermal cycle is the direct steam cycle. Steam from the geothermal well is passed through a turbine and exhausted to the atmosphere or to a condenser. Flash steam plants are used to generate power from liquid-dominated resources that are hot enough to flash a significant proportion of the water to steam in surface equipment, either at one or two pressure stages (single-flash or double-flash plants) as shown in Figs. 5.9 and 5.10, respectively. The steam flows through a steam turbine to produce power while the brine is reinjected back to the ground. Steam exiting the turbine is condensed with cooling water obtained in a cooling tower or a spray pond before being reinjected. Binary cycle plants use the geothermal brine from liquid-dominated resources (Fig. 5.11). These plants operate on a Rankine cycle with a binary

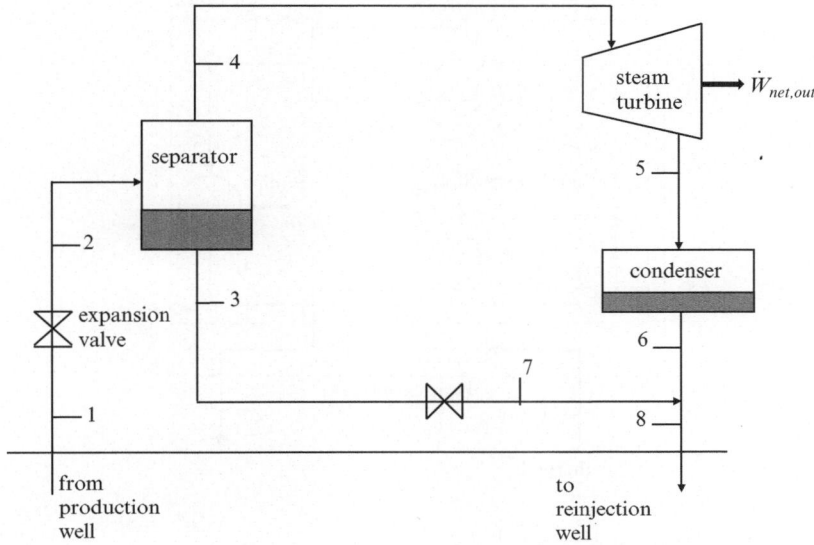

Fig. 5.9 Single-flash geothermal power plant

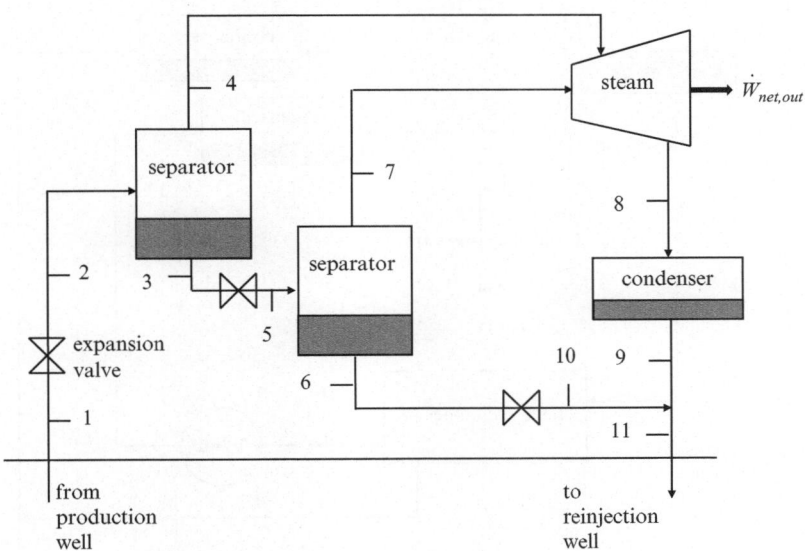

Fig. 5.10 Double-flash geothermal power plant

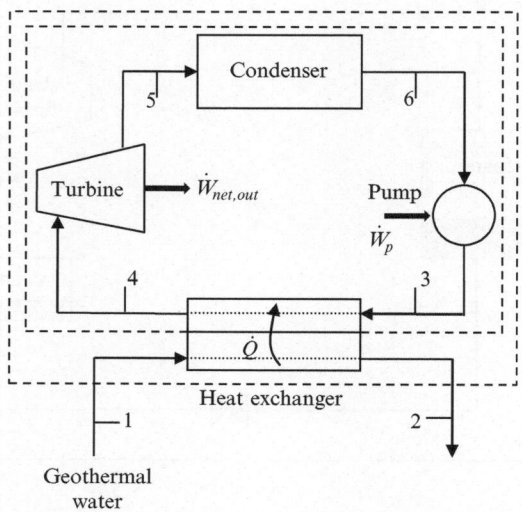

Fig. 5.11 Binary geothermal power plant

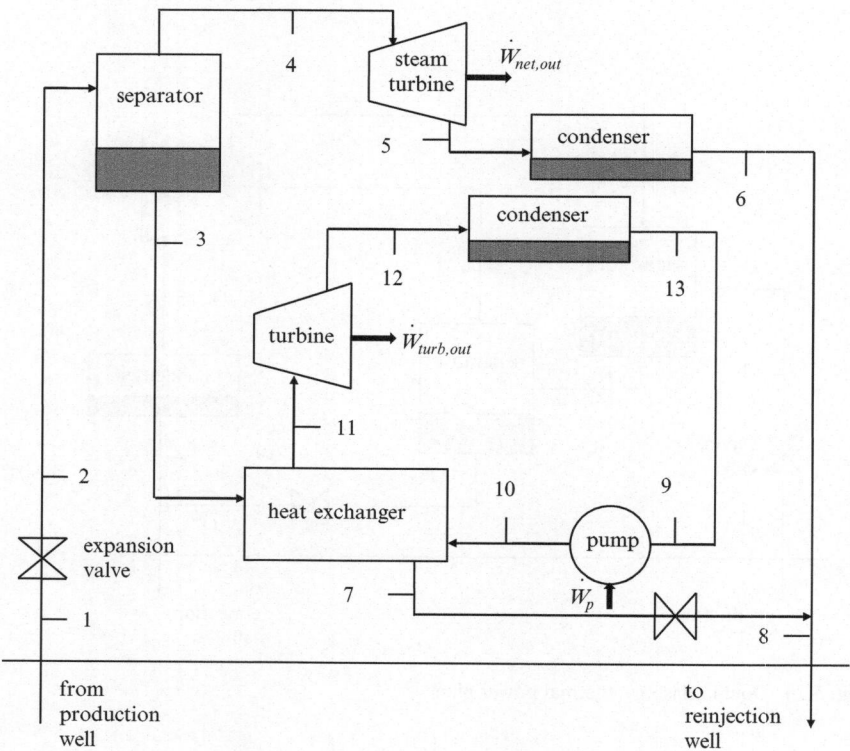

Fig. 5.12 Combined flash-binary geothermal power plant

working fluid (isobutane, isopentane, R-114, etc.) that has a low boiling temperature. The working fluid is completely vaporized and usually superheated by the geothermal heat in the vaporizer. The vapor expands in the turbine, and then condenses in a water-cooled condenser or dry cooling tower before being pumped back to the vaporizer to complete the cycle. Combined flash/binary plants (Fig. 5.12) incorporate both a binary unit and a flashing unit to exploit the advantages associated with both systems. The liquid portion of the geothermal mixture serves as the input heat for the binary cycle and the steam portion drives a steam turbine to produce power.

The actual heat input to a geothermal power cycle is less than the term in the denominator of (5.33) inasmuch as part of geothermal water is reinjected back to the ground at a temperature much greater than the temperature of the environment. In an approach that accounts for the actual reinjection temperature, the thermal efficiency is expressed as

$$\eta_{th-2} = \frac{\dot{W}_{net,\,out}}{\dot{Q}_{in}} \qquad (5.34)$$

For a single-flash cycle, the thermal efficiency may be expressed as

$$\eta_{th,\,single\,flash} = \frac{\dot{W}_{net,\,out}}{\dot{Q}_{in}} = \frac{\dot{W}_{net,\,out}}{\dot{m}_2 h_2 - \dot{m}_3 h_3} \qquad (5.35)$$

where the subscripts refer to state points in Fig. 5.9. For a double-flash cycle, the efficiency becomes

$$\eta_{th,\,double\,flash} = \frac{\dot{W}_{net,\,out}}{(\dot{m}_2 h_2 - \dot{m}_3 h_3) + (\dot{m}_5 h_5 - \dot{m}_6 h_6)} \qquad (5.36)$$

where the state points are shown in Fig. 5.10. Referring to Fig. 5.11, for a binary cycle we obtain

$$\eta_{th,\,binary} = \frac{\dot{W}_{net,\,out}}{\dot{m}_{geo}(h_1 - h_2)} \qquad (5.37)$$

or

$$\eta_{th,\,binary} = \frac{\dot{W}_{net,\,out}}{\dot{m}_{binary}(h_4 - h_3)} \qquad (5.38)$$

where $\dot{m}_{binary}$ is the mass flow rate of binary working fluid. For a combined flash-binary cycle, the thermal efficiency is

$$\eta_{\text{th, flash-binary}} = \frac{\dot{W}_{\text{net, out}}}{(\dot{m}_2 h_2 - \dot{m}_3 h_3) + (\dot{m}_3 h_3 - \dot{m}_7 h_7)} \tag{5.39}$$

where the state points are shown in Fig. 5.12.

Using the exergy of geothermal water (in the reservoir or at the well head) as the exergy input to the plant, the exergy efficiency of a geothermal power plant can be expressed as

$$\varepsilon = \frac{\dot{W}_{\text{net, out}}}{\dot{X}_{\text{in}}} = \frac{\dot{W}_{\text{net, out}}}{\dot{m}_{\text{geo}}\left[h_{\text{geo}} - h_0 - T_0\left(s_{\text{geo}} - s_0\right)\right]} \tag{5.40}$$

Using the exergy change of geothermal water in the cycle as the exergy input to the cycle, the exergy efficiencies may be expressed for single-flash, double-flash, and combined flash–binary cycles as

$$\varepsilon_{\text{single flash}} = \frac{\dot{W}_{\text{net, out}}}{\dot{m}_2 x_2 - \dot{m}_3 x_3}$$

$$= \frac{\dot{W}_{\text{net, out}}}{\dot{m}_2[h_2 - h_0 - T_0(s_2 - s_0)] - \dot{m}_3[h_3 - h_0 - T_0(s_3 - s_0)]} \tag{5.41}$$

$$\varepsilon_{\text{double flash}} = \frac{\dot{W}_{\text{net, out}}}{(\dot{m}_2 x_2 - \dot{m}_3 x_3) + (\dot{m}_5 x_5 - \dot{m}_6 x_6)} \tag{5.42}$$

$$\varepsilon_{\text{flash-binary}-1} = \frac{\dot{W}_{\text{net, out}}}{(\dot{m}_2 x_2 - \dot{m}_3 x_3) + (\dot{m}_3 x_3 - \dot{m}_7 x_7)} \tag{5.43}$$

where ex is the specific flow exergy of the fluid. For a binary cycle, the exergy efficiency may be defined based on the exergy decrease of geothermal water or the exergy increase of the binary working fluid in the heat exchanger. That is,

$$\varepsilon_{\text{binary}-1} = \frac{\dot{W}_{\text{net, out}}}{\dot{m}_{\text{geo}}[h_2 - h_1 - T_0(s_2 - s_1)]} \tag{5.44}$$

$$\varepsilon_{\text{binary}-2} = \frac{\dot{W}_{\text{net, out}}}{\dot{m}_{\text{binary}}[h_4 - h_3 - T_0(s_4 - s_3)]} \tag{5.45}$$

The difference between the denominators of (5.44) and (5.45) is the exergy destruction in the heat exchanger. The exergy efficiency definitions in (5.44) and (5.45) can be illustrated by considering the different systems indicated by the inner and outer dashed lines, respectively, in Fig. 5.11.

By adapting the approach used in (5.45), one may express the exergy efficiency for a combined flash–binary cycle as

$$\varepsilon_{\text{flash-binary-2}} = \frac{\dot{W}_{\text{net, out}}}{(\dot{m}_2 x_2 - \dot{m}_3 x_3) + \dot{m}_{\text{binary}}(x_{11} - x_{10})} \tag{5.46}$$

The efficiency in (5.43) is more advantageous than that in (5.46) because exergy input is expressed by the exergy change of geothermal water for both the flash and binary parts of the cycle in (5.43), respectively.

Example 5.3 Consider a binary geothermal power plant like that in Fig. 5.11 using geothermal water at 165°C with isobutane as the working fluid. The mass flow rate of geothermal water is 555 kg/s. In this cycle, isobutane is heated and vaporized in the heat exchanger by geothermal water. Then the isobutane flows through the turbine, is condensed, and pumped back to the heat exchanger, completing the binary cycle. The heat exchanger and condenser pressures are taken to be 3,000 and 400 kPa, respectively, and the temperature at the turbine inlet (or heat exchanger exit) is taken to be 150°C, which is 15°C lower than the geothermal water temperature at the heat exchanger inlet. The isentropic efficiencies of the turbine and pump are taken to be 80% and 70%, respectively. About 10% of the power output is used for internal demands such as powering fans in the air-cooled condenser. These values closely correspond to those of an actual power plant [31]. Noting that a pinch-point will occur at the start of vaporization of the working fluid in the heat exchanger, the energy balance relations for the heat exchanger can be written as

$$\dot{m}_{\text{geo}} c_{\text{geo}} \left[T_1 - (T_{\text{vap}} + \Delta T_{\text{pp}}) \right] = \dot{m}_{\text{binary}} \left(h_4 - h_{\text{binary},f} \right) \tag{5.47}$$

$$\dot{m}_{\text{geo}} c_{\text{geo}} \left[(T_{\text{vap}} + \Delta T_{\text{pp}}) - T_2 \right] = \dot{m}_{\text{binary}} \left(h_{\text{binary},f} - h_3 \right) \tag{5.48}$$

where $\dot{m}_{\text{binary}}$ is the mass flow rate of the binary fluid, c_{geo} is the specific heat of geothermal water, T_{vap} is the vaporization temperature of the binary fluid at the heat exchanger pressure, ΔT_{pp} is the pinch-point temperature difference, and $h_{\text{binary},f}$ is the specific enthalpy of the binary fluid at the start of vaporization. The pinch-point temperature difference is assumed to be 6°C. Equations 5.47 and 5.48 can be used to establish the mass flow rate of the binary fluid, and the geothermal water temperature at the heat exchanger exit. The analysis of the cycle with the stated values produces the following results.

$$\dot{W}_{\text{net, out}} = 22,382\,\text{kW}, \quad \dot{E}_{\text{in}} = 328,786\,\text{kW}, \quad \dot{Q}_{\text{in}} = 185,181\,\text{kW}, \quad T_2 = 86.6°\text{C}$$

$$\dot{X}_{\text{in}} = 60,014\,\text{kW}, \quad \Delta \dot{X}_{1-2} = 46,904\,\text{kW}, \quad \Delta \dot{X}_{3-4} = 37,316\,\text{kW},$$

$$n_{\text{th}-1} = 6.8\% \,[\text{Eq.}(5.33)], \quad \eta_{\text{th}-2} = 12.1\% \,[\text{Eq.}(5.37)]$$

$$\varepsilon = 37.3\% \,[\text{Eq.}(5.40)], \quad \varepsilon_{\text{binary}-1} = 47.7\% \,[\text{Eq.}(5.44)],$$
$$\varepsilon_{\text{binary}-2} = 60.0\% \,[\text{Eq.}(5.45)]$$

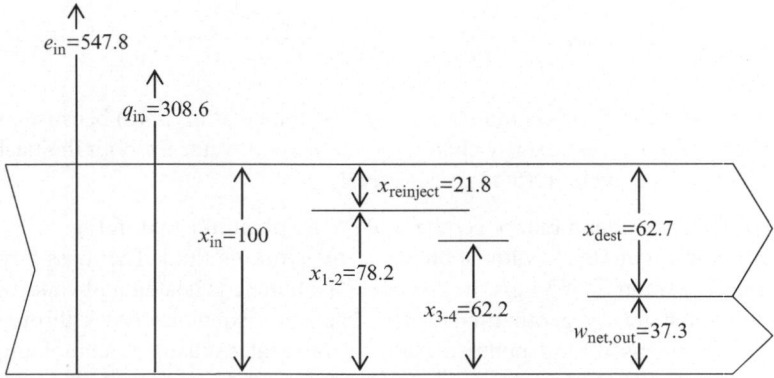

Fig. 5.13 Combined energy and exergy diagram for the binary geothermal power plant considered

It is clear that using different definitions leads to significantly different thermal and exergy efficiencies. This is typical of geothermal power plants. The results are presented in a combined energy and exergy diagram in Fig. 5.13. Because of the large range of values involved, the exergy of geothermal water at the heat exchanger inlet is normalized to 100 units of energy and other values are modified accordingly. The energy and exergy efficiencies can be obtained using terms in this diagram. The thermal and exergy efficiencies are 6.8% and 37.3%, respectively, based on the energy and exergy of geothermal water at the heat exchanger inlet. The thermal efficiency increases from 6.8% to 12.1% when the heat input to the binary fluid in the heat exchanger is used as the energy input to the cycle instead of the energy of the geothermal water at the inlet of the plant. This is analogous to using the heating value of fuel versus using the heat transferred to the steam in the boiler as the heat input to a steam power plant.

The exergy efficiency is only 37.3% when the exergy of geothermal water at the plant inlet is used. Using the exergy decrease of geothermal water in the heat exchanger as the exergy input to the cycle yields an exergy efficiency of 47.7% whereas using the exergy increase of the binary fluid yields an exergy efficiency of 60.0%. These three approaches are analogous to using the exergy of the fuel [(5.8)], the exergy transfer to the steam accompanying the heat input to the cycle [(5.7)], and the exergy increase of the steam in the boiler [(5.6)]. The exergy of the geothermal water at the exit of the heat exchanger, which is reinjected to the ground, represents 21.8% of the exergy input to the cycle. This significant percentage is due to the relatively high temperature of the geothermal water (86.6°C). The exergy destruction in the heat exchanger accounts for 16.0% of the exergy input. The remaining exergy destructions ($100 - 21.8 - 16.0 - 37.3 = 24.9$) are due to irreversibilities in the turbine, pump, and condenser.

The efficiencies for the plants considered as examples yield some important information on the relative magnitudes of heat losses and exergy destructions in the plants. Combined energy and exergy diagrams present the results concisely and clearly. The efficiencies for power cycles not specifically discussed in this chapter can be deduced from the relations given for the cycles considered.

For the current state of thermodynamics, it seems almost impossible to have a common efficiency definition for all energy systems. Therefore, the best way of avoiding misuse and misunderstanding is to define the efficiency used in any application carefully. An understanding of both energy and exergy efficiencies is essential for designing, analyzing, optimizing, and improving energy systems through appropriate energy policies and strategies. If such policies and strategies are in place, numerous measures can be applied to improve the efficiency of electrical generating plants. These measures should be weighed against other factors and, where appropriate, implemented. It should be understood that decisions on power plant operations are normally based primarily on economic criteria. Often other criteria such as environmental considerations are also important. Economic and exergy analyses can be combined by means of exergoeconomic analyses, which can include exergetic life cycle assessment. A rational efficiency definition should accompany such an analysis. It is more appropriate to use an exergy efficiency based on the exergy of the fuel in a fossil-fuel power or cogeneration system because an important part of exergy costing involves fuel cost. All exergy losses are accounted for in this approach. For renewable energy systems, it is more appropriate to use the exergy of an energy source as the exergy input to the system. This approach allows all exergy destructions to be accounted for, including those in heat exchange equipment. All losses are ultimately related to the economics of the system operation. The difference between efficiency definitions often relates to the selection of different system boundaries. Depending on the selection, losses occurring at a particular site may be accounted for in a definition or excluded [6].

A detailed case study on efficiency evaluation of a binary geothermal power plant is given in Appendix A.

5.6   Energetic and Exergetic Analyses of a Photovoltaic System

This analysis is based on references [32–34]. An example of solar technology is adopted to demonstrate the link between sustainability and efficiency. An effective way to maintain a good electrical efficiency by removing heat from the solar panels and to have a better overall efficiency of a photovoltaic system is to utilize both technologies simultaneously. This kind of system is known as a hybrid photovoltaic/thermal (PV/T) system and can be beneficial for low-temperature thermal applications such as water heating, air heating, agricultural crop drying, solar greenhouses, and space heating among others, along with electricity generation that can further be beneficial for rural electrification and agricultural applications including solar water pumping and so on. In this case study we give a simple demonstration of how both technologies together give better efficiency, which directly relates to better sustainability.

Based on the first law of thermodynamics, the energy efficiency of a PV/T system can be defined as a ratio of total energy (electrical and thermal) produced by a PV/T system to the total solar energy falling on the photovoltaic surface and can be given as [19, 20]:

$$\eta = \frac{E}{S_T A} = \frac{V_{oc} I_{sc} + \dot{Q}}{S_T A} \tag{5.49}$$

where

$$\dot{Q} = h_{ca} A (T_{cell} - T_{amb}) \text{ and } h_{ca} = 5.7 + 3.8 v.$$

Here, h_{ca}, A, T_{cell}, T_{amb}, I_{sc}, and V_{oc} are the convective heat transfer coefficient from the photovoltaic cell to the ambient area of the photovoltaic surface, cell temperature, ambient temperature, short circuit current, and open circuit voltage, respectively. The convective (and radiative) heat transfer coefficient from the photovoltaic cell to ambient, can be calculated by considering wind velocity (v), density of the air, and the surrounding (ambient) conditions.

The exergy efficiency is based on the second law of thermodynamics that not only gives the quantitative assessment of energy but also the qualitative. This is also called exergy efficiency. A comparison of the PV and PV/T systems is also presented in the form of a case study later in this section.

The exergy efficiency of a photovoltaic system can be given as

$$\varepsilon = \frac{\dot{X}}{\dot{X}_{solar}} \tag{5.50}$$

where $\dot{X}$ is the exergy of the PV system which is mainly the electrical power output of the system. The thermal energy gained by the system during operation is not desirable in the case of a PV system, therefore this becomes a heat loss to the system and hence needs to be subtracted from the former in order to calculate the exergy of a PV system. $\dot{X}_{solar}$ is the exergy rate from the solar irradiance in W/m² which can be given as

$$\dot{X}_{solar} = \left(1 - \frac{T_{amb}}{T_{sun}}\right) S_T A \tag{5.51}$$

An expression for the exergy of PV can be given as

$$\dot{X} = V_m I_m - \left(1 - \frac{T_{amb}}{T_{cell}}\right) \dot{Q} \tag{5.52}$$

Here, I_m and V_m are the actual current and voltage.

Unlike PV systems, the PV/T system uses the thermal energy available on the PV panel and this time the thermal energy gain can be utilized as useful energy and hence, the exergy of the PV/T system becomes the sum of the electrical exergy and thermal exergy of the system and the exergy efficiency can be defined as

$$\varepsilon = \frac{\dot{X}}{\dot{X}_{solar}} = \frac{\dot{X}_e + \dot{X}_{th}}{\dot{X}_{solar}} \tag{5.53}$$

An expression for the exergy of the PV/T system can be given as [20]:

$$\dot{X} = V_m I_m + \left(1 - \frac{T_{amb}}{T_{cell}}\right)\dot{Q} \tag{5.54}$$

We now apply the model presented above to some actual data sets as obtained through experiments in New Delhi, India, which is located at $77°12'E$ longitude and $28°35'N$ latitude. The test was performed from 9:00 a.m. to 4:00 p.m. on March 27, 2006 and the data measured included total solar irradiation, voltage, open-circuit voltage, current, short-circuit current, cell temperature, ambient temperature, and velocity of the air just above the photovoltaic surface. The data for hourly total solar radiation and the wind velocity were measured for different places on the photovoltaic surface and an average value for both was used to calculate the energy and exergy of the photovoltaic system. The uncertainty analysis of the measured global radiation was done and the internal estimate of uncertainty was evaluated following [32] and it was found that the value for uncertainty for the measured global radiation was 2.23%. The system included two modules in series, and the area of one solar cell is 0.0139 m^2. The number of solar cells in the two modules was 72. Therefore, the efficiency analysis of a PV system for its performance assessment is done here based on some experimental data as explained above (Fig. 5.14).

Using (5.47) through (5.52) and experimental data from Joshi, Dincer, and Reddy [33], energy and exergy efficiencies are calculated and shown in Fig. 5.13. It is clear from the figure that the energy efficiency (33–45%) is higher than that of the exergy efficiency (11–16%) of the PV/T system and (7.8–13.8%) of the PV system. Maximum exergy efficiency for the PV/T (16%) and PV (13.8) can also be seen at 4 p.m. where as a minimum exergy efficiency for the PV/T (11%) and PV (7.8%) is at 12 p.m. In the present study, natural air is used to derive the heat from the photovoltaic surface. However, if air is supplied beneath the photovoltaic surface by a forced mode (e.g., by putting a fan beneath the photovoltaic panel), more thermal energy can be removed in a better as well as convenient way. In that case a higher energy and exergy efficiency can be achieved.

Figure 5.15 shows the comparison of exergy efficiencies of both PV as well as PV/T systems. Comparing both curves one can see that the exergy efficiency of PV/T is on average 20% more than that of PV. Carbon dioxide, a major greenhouse gas, is responsible for global warming hence there is a need to understand the ways by which we can reduce such emissions. One solution to this problem can be adopting

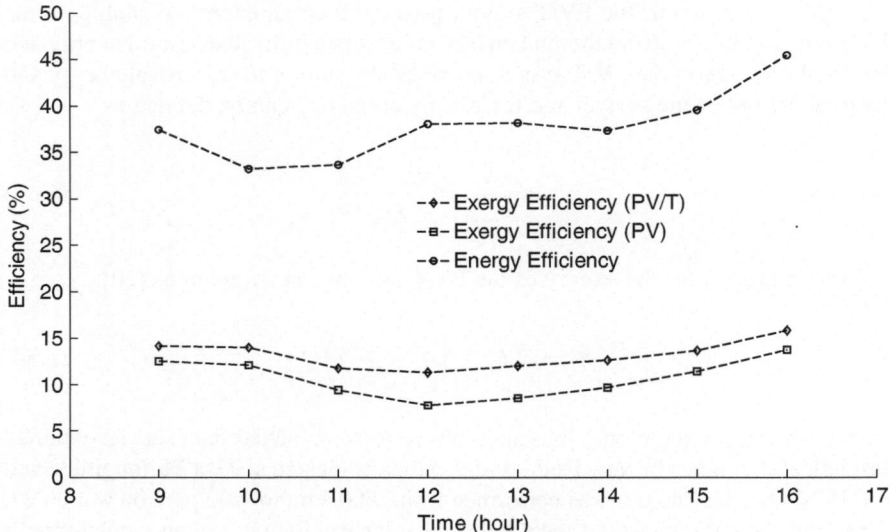

Fig. 5.14 Energy and exergy efficiencies of PV and PV/T systems (modified from Ref. [20])

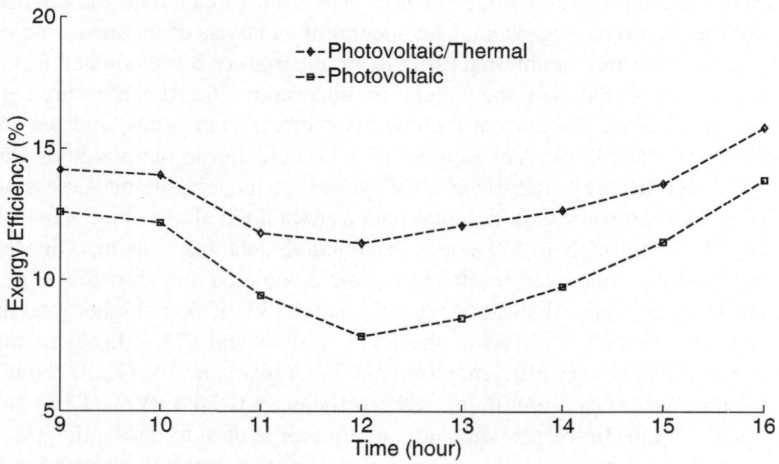

Fig. 5.15 Comparison of exergy efficiency of PV and PV/T systems (modified from Ref. [20])

unconventional energy sources wherever applicable; for example, for water heating one can use solar energy which is more eco-friendly as compared to using an electrical water heater that runs on electricity produced by conventional sources. Another example could be solar pumping: farmers irrigate their fields in the daytime and they can use solar water pumping instead of using an oil-based generator to produce electricity and use it to run the water pump. The unconventional energy sources (often called renewable energy sources) are environmentally benign as they emit fewer greenhouse gases into the atmosphere as compared to conventional ones. Although the nonrenewable or conventional sources of energy such as coal, oil, and natural gas are more economical than the renewable sources, they pollute the environment at a much faster rate. Coal-based electricity generation causes the highest greenhouse gas emission among all the conventional and unconventional sources during the operation and the installation of a power plant.

Chapter 6
Efficiencies of Refrigeration Systems

6.1 Refrigerators and Heat Pumps

A refrigerator is a device used to transfer heat from a low- to a high-temperature medium. They are cyclic devices. Figure 6.1a shows the schematic of a vapor-compression refrigeration cycle. A working fluid (called refrigerant) enters the compressor as a vapor and is compressed to the condenser pressure. The high-temperature refrigerant cools in the condenser by rejecting heat to a high-temperature medium (at T_H). The refrigerant enters the expansion valve as liquid. It is expanded in an expansion valve and its pressure and temperature drops. The refrigerant is a mixture of vapor and liquid at the inlet of the evaporator. It absorbs heat from a low-temperature medium (at T_L) as it flows in the evaporator. The cycle is completed when the refrigerant leaves the evaporator as a vapor and enters the compressor. The cycle is demonstrated in a simplified form in Fig. 6.1b.

An energy balance on the refrigeration cycle gives

$$Q_H = Q_L + W \qquad (6.1)$$

The efficiency indicator for a refrigeration cycle is the coefficient of performance (COP), which is defined as the heat absorbed from the cooled space divided by the work input in the compressor:

$$COP_R = \frac{Q_L}{W} \qquad (6.2)$$

This can also be expressed as

$$COP_R = \frac{Q_L}{Q_H - Q_L} = \frac{1}{Q_H/Q_L - 1} \qquad (6.3)$$

A heat pump is basically the same device as an evaporator. The difference is their purpose. The purpose of a refrigerator is to absorb heat from a cooled space to keep it at a desired low temperature (T_L). The purpose of a heat pump is to transfer

M. Kanoğlu et al., *Efficiency Evaluation of Energy Systems*,
SpringerBriefs in Energy, DOI 10.1007/978-1-4614-2242-6_6,

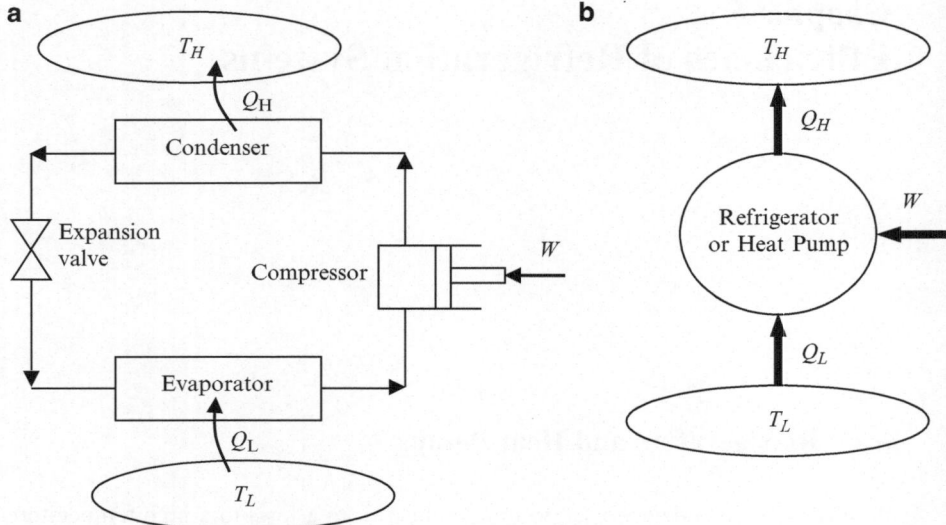

Fig. 6.1 (**a**) A vapor-compression refrigeration cycle. (**b**) Simplified schematic of refrigeration cycle

heat to a heated space to keep it at a desired high temperature (T_H). Thus, the COP of a heat pump is defined as

$$COP_{HP} = \frac{Q_H}{W} \tag{6.4}$$

This can also be expressed as

$$COP_{HP} = \frac{Q_H}{Q_H - Q_L} = \frac{1}{1 - Q_L/Q_H} \tag{6.5}$$

It can be easily shown that for given values Q_L and Q_H the COPs of a refrigerator and a heat pump are related to each other by

$$COP_{HP} = COP_R + 1 \tag{6.6}$$

This shows that the COP of a heat pump is greater than 1. The COP of a refrigerator can be less than or greater than 1.

6.1.1 Heat Pump Efficiencies

There are different criteria used to describe heat pump efficiency. In all of these criteria, the higher the number is, the higher the efficiency of the system. Heat pump efficiency is determined by comparing the amount of energy delivered by the heat pump to the amount of energy it consumes.

6.1.1.1 Coefficient of Performance (COP)

The COP is the most common measurement used to rate heat pump efficiency. COP is the ratio of the heat pump's heat output to the electrical energy input, as follows.

$$COP = \text{Heat Output/Electrical Energy Input} \tag{6.7}$$

6.1.1.2 Primary Energy Ratio (PER)

Heat pumps may be activated either electrically or by engines (e.g., internal combustion engines or gas motors). Unless electricity comes from an alternative source (e.g., hydro, wind, solar, etc.), heat pumps also utilize primary energy sources upstream like a thermoelectric plant or on-spot like a natural gas motor. When comparing heat pump systems driven by different energy sources it is more appropriate to use the PER, as defined by Holland et al. [35], as the ratio of useful heat delivered to the primary energy input. So this can be related to the COP by the following equation.

$$PER = \eta \, COP \tag{6.8}$$

where η is the efficiency with which the primary energy input is converted into work up to the shaft of the compressor.

However, due to high COP, the PER, as given below, becomes high relative to conventional fossil fuel fired systems. In the case of an electrically driven compressor where the electricity is generated from a coal-burning power plant, the efficiency η may be as low as 25%. The above equation indicates that gas engine driven heat pumps are very attractive from a primary energy ratio point of view because values for η of 0.75 or better can be obtained. However, heat recovery systems tend to be judged on their potential money savings rather than their potential energy savings.

6.1.1.3 Energy Efficiency Ratio (EER)

The EER is used for evaluating a heat pump's efficiency in the cooling cycle. The same rating system is used for air-conditioners, making it easy to compare different units. EER is the ratio of cooling capacity in Btu/h provided to electricity consumed in W as follows.

$$EER = \text{Cooling Capacity (Btu/h)/Electrical Energy Input(W)} \tag{6.9}$$

Because 1 W = 3.412 Btu/h, the relationship between the COP and EER is

$$EER = 3.412 \, COP \tag{6.10}$$

6.1.1.4 Heating Season Performance Factor (HSPF)

A heat pump's performance varies depending on the weather and how much supplementary heat is required. Therefore, a more realistic measurement, especially for air-to-air heat pumps, is calculated on a seasonal basis. These measurements are referred to as the heating season performance factor (HSPF) for the heating cycle. The industry standard test for overall heating efficiency provides an HSPF rating. Such a laboratory test attempts to take into account the reductions in efficiency caused by defrosting, temperature fluctuations, supplemental heat, fans, and on/off cycling. HSPF is the estimated seasonal heating output in Btu/h divided by the seasonal power consumption in W, as follows.

HSPF = Total Seasonal Heating Output (Btu/h)/Total Electrical Energy Input (W)   (6.11)

It can be thought of as the "average COP" for the entire heating system.

6.1.1.5 Seasonal Energy Efficiency Ratio (SEER)

As explained above, a heat pump's performance varies depending on the weather and the amount of supplementary heat required. Thus, a more realistic measurement, particularly for air-to-air heat pumps, is calculated on a seasonal basis. These measurements are referred to as the seasonal energy efficiency ratio (SEER) for the cooling cycle. Therefore SEER rates the seasonal cooling performance of the heat pump. The SEER is the ratio of the total cooling of the heat pump in Btu/h to the total electrical energy input in W during the same period.

SEER = Total Seasonal Cooling Output (Btu/h)/Total Electrical Energy Input (W)   (6.12)

Naturally, the SEER for a unit will vary depending on where in the country it is located. The higher the SEER is, the more efficiently the heat pump cools. The SEER is the ratio of heat energy removed from the house compared to the energy used to operate the heat pump, including fans. The SEER is usually noticeably higher than the HSPF because defrosting is not needed and there is no need for expensive supplemental heat during air-conditioning weather.

6.1.2   The Carnot Refrigeration Cycle

The Carnot cycle is a theoretical model that is useful for understanding a refrigeration cycle. It is the most efficient cycle operating between the given temperature limits T_H and T_L. The COP of the Carnot refrigeration cycle may be expressed as

$$\text{COP}_{R,\,rev} = \frac{Q_L}{W} = \frac{Q_L}{Q_H - Q_L} = \frac{T_L}{T_H - T_L} \qquad (6.13)$$

It can also be expressed as

$$\text{COP}_{R,\,rev} = \frac{1}{Q_H/Q_L - 1} = \frac{1}{T_H/T_L - 1} \qquad (6.14)$$

For a reversible heat pump, the following relations apply.

$$\text{COP}_{HP,\,rev} = \frac{Q_H}{W} = \frac{Q_H}{Q_H - Q_L} = \frac{T_H}{T_H - T_L} \qquad (6.15)$$

or

$$\text{COP}_{HP,\,rev} = \frac{1}{1 - Q_L/Q_H} = \frac{1}{1 - T_L/T_H} \qquad (6.16)$$

The above relations provide the maximum COPs for a refrigerator or a heat pump operating between the temperature limits of T_L and T_H. Actual refrigerators and heat pumps involve inefficiencies and thus, they will have lower COPs. The COP of a Carnot refrigeration cycle can be increased by either (a) increasing T_L or (b) decreasing T_H.

Example 6.1 A refrigeration cycle is used to keep a food department at $-15°C$ in an environment at $25°C$. The total heat gain to the food department is estimated to be 1,500 kJ/h and the heat rejection in the condenser is 2,600 kJ/h. Determine (a) the power input to the compressor in kW, (b) the COP of the refrigerator, and (c) the minimum power input to the compressor if a reversible refrigerator was used.

Solution

(a) The power input is determined from an energy balance on the refrigeration cycle:

$$\dot{W}_{in} = \dot{Q}_H - \dot{Q}_L = 2600 - 1500 = 1100\,\text{kJ/h} = (1100\,\text{kJ/h})\left(\frac{1\,\text{kW}}{3600\,\text{kJ/h}}\right)$$

$$= 0.306\,\text{kW}$$

(b) The COP of the refrigerator is

$$\text{COP}_R = \frac{\dot{Q}_L}{\dot{W}_{in}} = \frac{(1500/3600)\text{kW}}{0.306\,\text{kW}} = 1.36$$

(c) The maximum COP of the cycle and the corresponding minimum power input are

$$\text{COP}_{R,\,rev} = \frac{T_L}{T_H - T_L} = \frac{258}{298 - 258} = 6.45$$

$$\dot{W}_{min} = \frac{\dot{Q}_L}{\text{COP}_{R,\,rev}} = \frac{(1500/3600)\text{kW}}{6.45} = \mathbf{0.065\,kW}$$

6.2   Second Law Analysis of Vapor-Compression Refrigeration Cycle

Consider the vapor-compression refrigeration cycle operating between a low-temperature medium at T_L and a high-temperature medium at T_H as shown in Fig. 6.2. Actual refrigeration cycles are not as efficient as ideal ones such as the Carnot cycle because of the irreversibilities involved. But the conclusion we can draw from (6.13) that the COP is inversely proportional to the temperature difference $T_H - T_L$ is equally valid for actual refrigeration cycles.

The goal of a second law or exergy analysis of a refrigeration system is to determine the components that can benefit the most by improvements. This is done by identifying the locations of greatest exergy destruction and the components with the lowest exergy or second law efficiency. Exergy destruction in a component can be determined directly from an exergy balance or indirectly by first calculating the entropy generation and then using the relation

$$\dot{X}_{dest} = T_0 \dot{S}_{gen} \tag{6.17}$$

where T_0 is the environment (the dead-state) temperature. For a refrigerator, T_0 is usually the temperature of the high-temperature medium T_H (for a heat pump it is T_L). Exergy destructions and exergy or the second law efficiencies for major components of a refrigeration system operating on the cycle shown in Fig. 6.2 may be written as follows.

Compressor (adiabatic):

$$\dot{X}_{dest,1-2} = T_0 \dot{S}_{gen,1-2} = \dot{m} T_0 (s_2 - s_1) \tag{6.18}$$

$$\varepsilon_{Comp} = \frac{\dot{X}_{recovered}}{\dot{X}_{expended}} = \frac{\dot{W}_{rev}}{\dot{W}_{act,\,in}} = \frac{\dot{m}[h_2 - h_1 - T_0(s_2 - s_1)]}{\dot{m}(h_2 - h_1)} = \frac{\psi_2 - \psi_1}{h_2 - h_1}$$

$$= 1 - \frac{\dot{X}_{dest,\,1-2}}{\dot{W}_{act,\,in}} \tag{6.19}$$

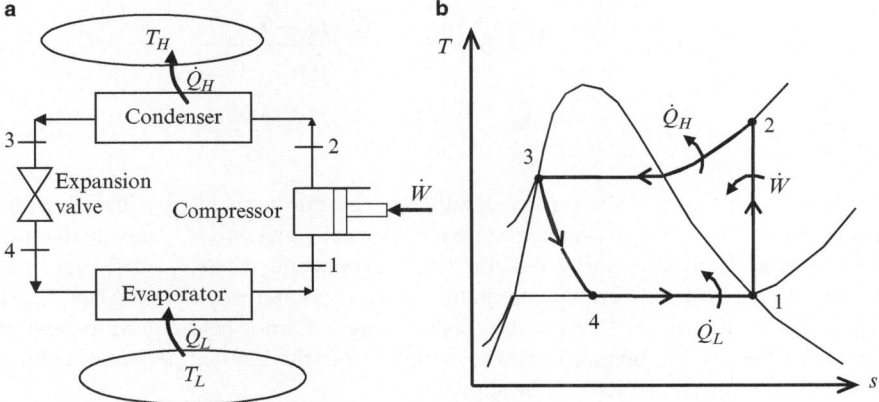

Fig. 6.2 A vapor-compression refrigeration system for analysis and its temperature-entropy diagram for the ideal case

Condenser:

$$\dot{X}_{dest,\,2-3} = T_0 \dot{S}_{gen,\,2-3} = T_0 \left[\dot{m}(s_3 - s_2) + \frac{\dot{Q}_H}{T_H} \right] \tag{6.20}$$

$$\varepsilon_{Cond} = \frac{\dot{X}_{recovered}}{\dot{X}_{expended}} = \frac{\dot{X}_{\dot{Q}_H}}{\dot{X}_2 - \dot{X}_3} = \frac{\dot{Q}_H(1 - T_0/T_H)}{\dot{X}_2 - X_3}$$

$$= \frac{\dot{Q}_H(1 - T_0/T_H)}{\dot{m}[h_2 - h_3 - T_0(s_2 - s_3)]} = 1 - \frac{\dot{X}_{dest,2-3}}{\dot{X}_2 - \dot{X}_3} \tag{6.21}$$

Note that when $T_H = T_0$, which is usually the case for refrigerators, $\varepsilon_{Cond} = 0$ because there is no recoverable exergy in this case.

Expansion valve:

$$\dot{X}_{dest,\,3-4} = T_0 \dot{S}_{gen,\,3-4} = \dot{m} T_0 (s_4 - s_3) \tag{6.22}$$

$$\varepsilon_{ExpValve} = \frac{\dot{X}_{recovered}}{\dot{X}_{expended}} = \frac{0}{\dot{X}_3 - \dot{X}_4} = 0 \quad \text{or}$$

$$\varepsilon_{ExpValve} = 1 - \frac{\dot{X}_{dest,\,3-4}}{\dot{X}_{expended}} = 1 - \frac{\dot{X}_3 - \dot{X}_4}{\dot{X}_3 - \dot{X}_4} = 0 \tag{6.23}$$

Evaporator:

$$\dot{X}_{dest,\,4-1} = T_0 \dot{S}_{gen,\,4-1} = T_0 \left[\dot{m}(s_1 - s_4) - \frac{\dot{Q}_L}{T_L} \right] \tag{6.24}$$

$$\varepsilon_{\text{Evap}} = \frac{\dot{X}_{\text{recovered}}}{\dot{X}_{\text{expended}}} = \frac{\dot{X}_{\dot{Q}_L}}{\dot{X}_4 - \dot{X}_1} = \frac{\dot{Q}_L(T_0 - T_L)/T_L}{\dot{X}_4 - \dot{X}_1}$$

$$= \frac{\dot{Q}_L(T_0 - T_L)/T_L}{\dot{m}[h_4 - h_1 - T_0(s_4 - s_1)]} = 1 - \frac{\dot{X}_{\text{dest, }4-1}}{\dot{X}_4 - \dot{X}_1} \qquad (6.25)$$

Here $\dot{X}_{\dot{Q}_L}$ represents the positive of the exergy rate associated with the withdrawal of heat from the low-temperature medium at T_L at a rate of $\dot{Q}_L$. Note that the directions of heat and exergy transfer become opposite when $T_L < T_0$ (i.e., the exergy of the low-temperature medium increases as it loses heat). Also, $\dot{X}_{\dot{Q}_L}$ is equivalent to the power that can be produced by a Carnot heat engine receiving heat from the environment at T_0 and rejecting heat to the low-temperature medium at T_L at a rate of $\dot{Q}_L$, which can be shown to be

$$\dot{X}_{\dot{Q}_L} = \dot{Q}_L \frac{T_0 - T_L}{T_L} \qquad (6.26)$$

From the definition of reversibility, this is equivalent to the minimum or reversible power input required to remove heat at a rate of $\dot{Q}_L$ and reject it to the environment at T_0. That is, $\dot{W}_{rev, \text{in}} = \dot{W}_{\text{min, in}} = \dot{X}_{\dot{Q}_L}$.

Note that when $T_L = T_0$, which is often the case for heat pumps, $\eta_{\text{II, Evap}} = 0$ inasmuch as there is no recoverable exergy in this case.

The total exergy destruction associated with the cycle is the sum of the exergy destructions:

$$\dot{X}_{\text{dest, total}} = \dot{X}_{\text{dest, }1-2} + \dot{X}_{\text{dest, }2-3} + \dot{X}_{\text{dest, }3-4} + \dot{X}_{\text{dest, }4-1} \qquad (6.27)$$

It can be shown that the total exergy destruction associated with a refrigeration cycle can also be obtained by taking the difference between the exergy supplied (power input) and the exergy recovered (the exergy of the heat withdrawn from the low-temperature medium):

$$\dot{X}_{\text{dest, total}} = \dot{W}_{\text{in}} - \dot{X}_{\dot{Q}_L} \qquad (6.28)$$

The second law or exergy efficiency of the cycle can then be expressed as

$$\varepsilon_{\text{cycle}} = \frac{\dot{X}_{\dot{Q}_L}}{\dot{W}_{\text{in}}} = \frac{\dot{W}_{\text{min, in}}}{\dot{W}_{\text{in}}} = 1 - \frac{\dot{X}_{\text{dest, total}}}{\dot{W}_{\text{in}}} \qquad (6.29)$$

Substituting $\dot{W}_{\text{in}} = \dfrac{\dot{Q}_L}{\text{COP}_R}$ and $\dot{X}_{\dot{Q}_L} = \dot{Q}_L \dfrac{T_0 - T_L}{T_L}$ into (6.29) gives

$$\varepsilon_{\text{cycle}} = \frac{\dot{X}_{\dot{Q}_L}}{\dot{W}_{\text{in}}} = \frac{\dot{Q}_L(T_0 - T_L)/T_L}{\dot{Q}_L/\text{COP}_R} = \frac{\text{COP}_R}{T_L/(T_H - T_L)} = \frac{\text{COP}_R}{\text{COP}_{R, \text{rev}}} \qquad (6.30)$$

Fig. 6.3 Temperature-entropy diagram of vapor-compression refrigeration cycle considered in Example 6.2

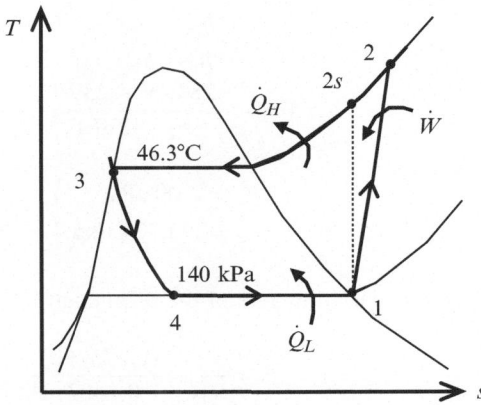

because $T_0 = T_H$ for a refrigeration cycle. Thus, the second law efficiency is also equal to the ratio of actual and maximum COPs for the cycle. This second law efficiency definition accounts for all irreversibilities associated within the refrigerator, including the heat transfers with the refrigerated space and the environment.

Example 6.2   A refrigerator using R-134a as the refrigerant is used to keep a space at $-10°C$ by rejecting heat to ambient air at $22°C$. R-134a enters the compressor at 140 kPa at a flow rate of 375 L/min as a saturated vapor. The isentropic efficiency of the compressor is 80%. The refrigerant leaves the condenser at $46.3°C$ as a saturated liquid. Determine (a) the rate of cooling provided by the system, (b) the COP, (c) the exergy destruction in each component of the cycle, (d) the second law efficiency of the cycle, and (e) the total exergy destruction in the cycle.

Solution   The temperature-entropy diagram of the cycle is given in Fig. 6.3.

(a) The properties of R-134a are (R-134a tables)

$$P_1 = \quad 140\,\text{kPa} \left.\begin{array}{l} \\ \\ \\ \end{array}\right\} \begin{array}{l} h_1 = 239.17\,\text{kJ/kg} \\ s_1 = 0.9446\,\text{kJ/kg} \cdot \text{K} \\ v_1 = 0.1402\,\text{m}^3/\text{kg} \end{array}$$

$$x_1 = \quad 1$$

$$P_3 = P_{\text{sat@46.3°C}} = 1200\,\text{kPa}$$

$$P_2 = 1200\,\text{kPa} \left.\begin{array}{l} \\ \\ \end{array}\right\} h_{2s} = 284.09\,\text{kJ/kg}$$

$$s_2 = s_1 = 0.9446\,\text{kJ/kg} \cdot \text{K}$$

$$P_3 = 1200\,\text{kPa} \left.\begin{array}{l} \\ \\ \end{array}\right\} \begin{array}{l} h_3 = 117.77\,\text{kJ/kg} \\ s_3 = 0.4244\,\text{kJ/kg} \cdot \text{K} \end{array}$$

$$x_3 = 0$$

$$h_4 = h_3 = 117.77\,\text{kJ/kg}$$

$$P_4 = 140\,\text{kPa} \left.\begin{array}{l} \\ \\ \end{array}\right\} s_4 = 0.4674\,\text{kJ/kg} \cdot \text{K}$$

$$h_4 = 117.77\,\text{kJ/kg}$$

$$\eta_C = \frac{h_{2s} - h_1}{h_2 - h_1}$$

$$0.80 = \frac{284.09 - 239.17}{h_2 - 239.17} \longrightarrow h_2 = 295.32 \,\text{kJ/kg}$$

$$\left.\begin{array}{l} P_2 = 1200 \,\text{kPa} \\ h_2 = 295.32 \,\text{kJ/kg} \end{array}\right\} s_2 = 0.9783 \,\text{kJ/kg} \cdot \text{K}$$

The mass flow rate of the refrigerant is

$$\dot{m} = \frac{\dot{V}_1}{v_1} = \frac{(0.375/60) \,\text{m}^3/s}{0.1402 \,\text{m}^3/\text{kg}} = 0.04458 \,\text{kg/s}$$

The refrigeration load, the rate of heat rejected, and the power input are

$$\dot{Q}_L = \dot{m}(h_1 - h_4) = (0.04458 \,\text{kg/s})(239.17 - 117.77)\text{kJ/kg} = \mathbf{5.41 \,kW}$$

$$\dot{Q}_H = \dot{m}(h_2 - h_3) = (0.04458 \,\text{kg/s})(295.32 - 117.77)\text{kJ/kg} = 7.92 \,\text{kW}$$

$$\dot{W} = \dot{m}(h_2 - h_1) = (0.04458 \,\text{kg/s})(295.32 - 239.17)\text{kJ/kg} = 2.50 \,\text{kW}$$

(b) The COP of the cycle is

$$\text{COP} = \frac{\dot{Q}_L}{\dot{W}_{in}} = \frac{5.41 \,\text{kW}}{2.50 \,\text{kW}} = \mathbf{2.16}$$

(c) Noting that the dead-state temperature is $T_0 = T_H = 295 \,\text{K}$, the exergy destruction in each component of the cycle is determined as follows.

Compressor:

$$\dot{S}_{gen, \, 1-2} = \dot{m}(s_2 - s_1) = (0.04458 \,\text{kg/s})(0.9783 - 0.9446)\text{kJ/kg} \cdot \text{K}$$
$$= 0.001502 \,\text{kW/K}$$

$$\dot{X}_{dest, \, 1-2} = T_0 \dot{S}_{gen, \, 1-2} = (295 \,\text{K})(0.001502 \,\text{kW/K}) = \mathbf{0.4432 \,kW}$$

Condenser:

$$\dot{S}_{gen, \, 2-3} = \dot{m}(s_3 - s_2) + \frac{\dot{Q}_H}{T_H}$$

$$= (0.04458 \,\text{kg/s})(0.4244 - 0.9783)\text{kJ/kg} \cdot \text{K} + \frac{7.92 \,\text{kW}}{295 \,\text{K}} = 0.002138 \,\text{kW/K}$$

$$\dot{X}_{\text{dest, 2-3}} = T_0\dot{S}_{\text{gen, 2-3}} = (295\,\text{K})(0.002138\,\text{kJ/kg} \cdot \text{K}) = \mathbf{0.6308\,kW}$$

Expansion valve:

$$\dot{S}_{\text{gen, 3-4}} = \dot{m}(s_4 - s_3) = (0.04458\,\text{kg/s})(0.4674 - 0.4244)\text{kJ/kg} \cdot \text{K}$$
$$= 0.001916\,\text{kW/K}$$

$$\dot{X}_{\text{dest, 3-4}} = T_0\dot{S}_{\text{gen, 3-4}} = (295\,\text{K})(0.001916\,\text{kJ/kg} \cdot \text{K}) = \mathbf{0.5651\,kW}$$

Evaporator:

$$\dot{S}_{\text{gen, 4-1}} = \dot{m}(s_1 - s_4) - \frac{\dot{Q}_L}{T_L}$$
$$= (0.04458\,\text{kg/s})(0.9446 - 0.4674)\text{kJ/kg} \cdot \text{K} - \frac{5.41\,\text{kW}}{263\,\text{K}} = 0.0006964\,\text{kW/K}$$

$$\dot{X}_{\text{dest, 4-1}} = T_0\dot{S}_{\text{gen, 4-1}} = (295\,\text{K})(0.0006964\,\text{kW/K}) = \mathbf{0.2054\,kW}$$

(d) The exergy of the heat transferred from the low-temperature medium is

$$\dot{Ex}_{\dot{Q}_L} = -\dot{Q}_L\left(1 - \frac{T_0}{T_L}\right) = -(5.41\,\text{kW})\left(1 - \frac{295}{263}\right) = 0.3163\,\text{kW}$$

This is also the minimum power input for the cycle. The second law efficiency of the cycle is

$$\varepsilon = \frac{\dot{Ex}_{\dot{Q}_L}}{\dot{W}} = \frac{0.3163}{2.503} = \mathbf{0.263 \text{ or } 26.3\%}$$

This efficiency may also be determined from

$$\varepsilon = \frac{\text{COP}}{\text{COP}_{\text{rev}}}$$

where

$$\text{COP}_{\text{R, rev}} = \frac{T_L}{T_H - T_L} = \frac{(-10 + 273)\text{K}}{[22 - (-10)]\text{K}} = 8.22$$

Substituting,

$$\varepsilon = \frac{\text{COP}}{\text{COP}_{\text{rev}}} = \frac{2.16}{8.22} = 0.263 \text{ or } 26.3\%$$

The results are identical as expected.

(e) The total exergy destruction in the cycle is the difference between the exergy supplied (power input) and the exergy recovered (the exergy of the heat transferred from the low-temperature medium):

$$\dot{X}_{\text{dest, total}} = \dot{W} - \dot{Ex}_{\dot{Q}_L} = 2.503 - 0.3163 = \mathbf{1.845\,kW}$$

The total exergy destruction can also be determined by adding exergy destructions in each component:

$$\dot{X}_{\text{dest, total}} = \dot{X}_{\text{dest, 1-2}} + \dot{X}_{\text{dest, 2-3}} + \dot{X}_{\text{dest, 3-4}} + \dot{X}_{\text{dest, 4-1}}$$
$$\doteq 0.4432 + 0.6308 + 0.5651 + 0.2054 = 1.845\,kW$$

The results are identical as expected. The exergy input to the cycle is equal to the actual work input, which is 2.503 kW. The same cooling load could have been accomplished by only 26.3% of this power (0.3163 kW) if a reversible system were used. The difference between the two is the exergy destroyed in the cycle (1.845 kW). It can be shown that increasing the evaporating temperature and decreasing the condensing temperature would also decrease the exergy destruction in these components.

6.3   Energy and Exergy Efficiencies of Vapor-Compression Heat Pump Cycle

Energy and exergy analyses of a vapor-compression heat pump cycle are very similar to the energy analysis of the vapor-compression refrigeration cycle as given earlier in this chapter. We refer to Fig. 6.2 in the following treatment. A heat pump is used to supply heat to the high-temperature space. Therefore, the coefficient of performance of the heat pump cycle is defined as

$$\text{COP} = \frac{\dot{Q}_H}{\dot{W}} \tag{6.31}$$

The maximum COP of a heat pump cycle operating between temperature limits of T_L and T_H based on the Carnot heat pump cycle was given as

$$\text{COP}_{\text{Carnot}} = \frac{T_H}{T_H - T_L} = \frac{1}{1 - T_L/T_H} \tag{6.32}$$

This is the maximum COP a heat pump operating between T_L and T_H can have. Equation 6.13 indicates that a smaller temperature difference between the heat sink and the heat source $(T_H - T_L)$ provides greater heat pump COP.

The aim in an exergy analysis is usually to determine the exergy destruction in each component of the system and to determine exergy efficiencies. The components with greater exergy destruction are also those with more potential for improvements. Exergy destruction in a component can be determined from an exergy balance on the component. It can also be determined by first calculating the entropy generation and using

$$\dot{Ex}_{\text{dest}} = T_0 \dot{S}_{\text{gen}} \tag{6.33}$$

where T_0 is the dead-state temperature or environment temperature. In a heat pump, T_0 is usually equal to the temperature of the low-temperature medium T_L. Exergy destruction and exergy efficciencies for compressor and expansion valve are the same as those in a refrigeration cycle as discussed in Sect. 6.2. For condenser and evaporator, we have

Condenser:

$$\dot{X}_{\text{dest, 2-3}} = T_0 \dot{S}_{\text{gen, 2-3}} = \dot{m} T_0 \left(s_3 - s_2 + \frac{q_H}{T_H} \right) \tag{6.34}$$

$$\varepsilon_{\text{Cond}} = \frac{\dot{X}_{\text{recovered}}}{\dot{X}_{\text{expended}}} = \frac{\dot{X}_{\dot{Q}_H}}{\dot{X}_2 - \dot{X}_3} = \frac{\dot{Q}_H (1 - T_0/T_H)}{\dot{X}_2 - \dot{X}_3}$$

$$= \frac{\dot{Q}_H (1 - T_0/T_H)}{\dot{m}[h_2 - h_3 - T_0(s_2 - s_3)]} = 1 - \frac{\dot{X}_{\text{dest, 2-3}}}{\dot{X}_2 - \dot{X}_3} \tag{6.35}$$

Evaporator:

$$\dot{Ex}_{\text{dest, 4-1}} = T_0 \dot{S}_{\text{gen, 4-1}} = \dot{m} T_0 \left(s_1 - s_4 - \frac{q_L}{T_L} \right) \tag{6.36}$$

$$\varepsilon_{\text{Evap}} = \frac{\dot{X}_{\text{recovered}}}{\dot{X}_{\text{expended}}} = \frac{\dot{X}_{\dot{Q}_L}}{\dot{X}_4 - \dot{X}_1} = \frac{\dot{Q}_L (T_0 - T_L)/T_L}{\dot{X}_4 - \dot{X}_1}$$

$$= \frac{\dot{Q}_L (T_0 - T_L)/T_L}{\dot{m}[h_4 - h_1 - T_0(s_4 - s_1)]} = 1 - \frac{\dot{X}_{\text{dest,4-1}}}{\dot{X}_4 - \dot{X}_1} \tag{6.37}$$

Note that when $T_L = T_0$, which is usually the case for heat pumps, $e_{\text{Evap}} = 0$ because there is no recoverable exergy in this case.

The total exergy destruction in the cycle can be determined by adding the exergy destruction in each component:

$$\dot{X}_{\text{dest, total}} = \dot{X}_{\text{dest, 1-2}} + \dot{X}_{\text{dest, 2-3}} + \dot{X}_{\text{dest, 3-4}} + \dot{X}_{\text{dest, 4-1}} \tag{6.38}$$

It can be shown that the total exergy destruction in the cycle can also be expressed as the difference between the exergy supplied (power input) and the exergy recovered (the exergy of the heat transferred to the high-temperature medium):

$$\dot{X}_{\text{dest, total}} = \dot{W} - \dot{X}_{\dot{Q}_H} \tag{6.39}$$

where the exergy of the heat transferred to the high-temperature medium is given by

$$\dot{X}_{\dot{Q}_H} = \dot{Q}_H \left(1 - \frac{T_0}{T_H}\right) \tag{6.40}$$

This is in fact the minimum power input to accomplish the required heating load $\dot{Q}_H$:

$$\dot{W}_{min} = \dot{X}_{\dot{Q}_H} \tag{6.41}$$

The second law efficiency (or exergy efficiency) of the cycle is defined as

$$\varepsilon = \frac{\dot{X}_{\dot{Q}_H}}{\dot{W}} = \frac{\dot{W}_{min}}{\dot{W}} = 1 - \frac{\dot{X}_{dest, total}}{\dot{W}} \tag{6.42}$$

Substituting

$$\dot{W} = \frac{\dot{Q}_H}{COP} \quad \text{and} \quad \dot{X}_{\dot{Q}_H} = \dot{Q}_H \left(1 - \frac{T_0}{T_H}\right)$$

into the second law efficiency equation

$$\varepsilon = \frac{\dot{X}_{\dot{Q}_H}}{\dot{W}} = \frac{\dot{Q}_H \left(1 - \frac{T_0}{T_H}\right)}{\frac{\dot{Q}_H}{COP}} = \dot{Q}_H \left(1 - \frac{T_0}{T_H}\right) \frac{COP}{\dot{Q}_H} = \frac{COP}{\frac{T_H}{T_H - T_L}} = \frac{COP}{COP_{Carnot}} \tag{6.43}$$

because $T_0 = T_L$. Thus, the second law efficiency is also equal to the ratio of actual and maximum COPs for the cycle. This second law efficiency definition accounts for irreversibilities within the heat pump inasmuch as heat transfers with the high- and low-temperature reservoirs are assumed reversible.

Example 6.3 A heat pump is used to keep a room at 25°C by rejecting heat to an environment at 5°C. The total heat loss from the room to the environment is estimated to be 45,000 kJ/h and the power input to the compressor is 6.5 kW. Determine (a) the rate of heat absorbed from the environment in kJ/h, (b) the COP of the heat pump, (c) the maximum rate of heat supply to the room for the given power input, and (d) the second law efficiency of the cycle. (e) Also, determine the minimum power input for the same heating load and the exergy destruction of the cycle.

Solution

(a) The rate of heat absorbed from the environment in kJ/h is

$$\dot{Q}_L = \dot{Q}_H - \dot{W} = 45,000 \, kJ/h - (4.5 \, kW) \left(\frac{3600 \, kJ/h}{1 \, kW}\right) = \mathbf{28,800 \, kJ/h}$$

(b) The COP of the heat pump is

$$\text{COP} = \frac{\dot{Q}_H}{\dot{W}} = \frac{(45,000/3600)\text{kW}}{4.5\,\text{kW}} = \mathbf{2.78}$$

(c) The COP of the Carnot cycle operating between the same temperature limits and the maximum rate of heat supply to the room for the given power input are

$$\text{COP}_{\text{Carnot}} = \frac{T_H}{T_H - T_L} = \frac{298}{298 - 278} = 14.9$$

$$\dot{Q}_{H,\,\text{max}} = \dot{W}\text{COP}_{\text{Carnot}} = (4.5\,\text{kW})\left(\frac{3600\,\text{kJ/h}}{1\,\text{kW}}\right)(14.9) = \mathbf{241,380\,kJ/h}$$

(d) The second law efficiency of the cycle is

$$\varepsilon = \frac{\text{COP}}{\text{COP}_{\text{carnot}}} = \frac{2.78}{14.9} = 0.186 \text{ or } \mathbf{18.6\%}$$

(e) The minimum power input for the same heating load and the exergy destruction of the cycle are

$$\dot{W}_{\text{min}} = \dot{X}_{\dot{Q}_H} = \dot{Q}_H\left(1 - \frac{T_0}{T_H}\right) = (45,000\,\text{kJ/h})\left(1 - \frac{278}{298}\right) = \mathbf{3020\,kJ/h}$$

$$\dot{X}_{\text{dest}} = \dot{W} - \dot{W}_{\text{min}} = (4.5 \times 3600)\text{kJ/h} - 3020\,\text{kJ/h} = \mathbf{13,180\,kJ/h}$$

The second law efficiency may alternatively be determined from

$$\varepsilon = \frac{\dot{W}_{\text{min}}}{\dot{W}} = \frac{3020\,\text{kJ/h}}{(4.5 \times 3600)\text{kJ/h}} = 0.186 \text{ or } 18.6\%$$

The result is the same as expected.

6.4 Absorption Refrigeration Cycle

An absorption refrigeration cycle uses a heat source in place of power input to a compressor (Fig. 6.4). The system involves a power input to a pump and heat input in a generator. Therefore, the coefficient of performance of the system can be expressed as

$$\text{COP} = \frac{\dot{Q}_L}{\dot{W}_P + \dot{Q}_{\text{gen}}} \tag{6.44}$$

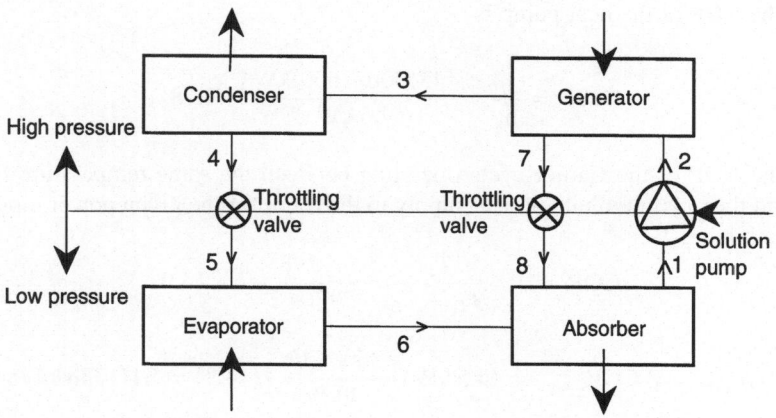

Fig. 6.4 The basic absorption refrigeration system

where $\dot{Q}_L$ is the rate of cooling in the evaporator, $\dot{Q}_{gen}$ is the rate of heat supplied to the generator, and $\dot{W}_P$ is the power input to the pump. The power input is usually neglected as its value is small compared to heat input. The second law efficiency may be expressed as

$$\eta_{II} = \frac{\dot{X}_{\dot{Q}_L}}{\dot{W}_P + \dot{X}_{\dot{Q}_{gen}}} = \frac{-\dot{Q}_L(1 - T_0/T_L)}{\dot{W}_P + \dot{Q}_H(1 - T_0/T_s)} \tag{6.45}$$

where T_0, T_L, and T_s are the temperatures of dead-state, cooled space, and heat source, respectively.

In order to develop a relation for the maximum (reversible) COP of an absorption refrigeration system, we consider a reversible heat engine and a reversible refrigerator as shown in Fig. 6.5. Heat is absorbed from a source at T_s by a reversible heat engine and the waste heat is rejected to an environment T_0. Work output from the heat engine is used as the work input in the reversible refrigerator, which keeps a refrigerated space at T_L while rejecting heat to the environment at T_0. Using the definition of COP for an absorption refrigeration system, thermal efficiency of a reversible heat engine, and the COP of a reversible refrigerator, we obtain

$$\text{COP}_{abs, rev} = \frac{\dot{Q}_L}{\dot{Q}_{gen}} = \frac{\dot{W}}{\dot{Q}_{gen}}\frac{\dot{Q}_L}{\dot{W}} = \eta_{th, rev}\text{COP}_{R, rev} = \left(1 - \frac{T_0}{T_s}\right)\left(\frac{T_L}{T_0 - T_L}\right) \tag{6.46}$$

Using this result, the second law efficiency of an absorption refrigeration system can also be expressed as

$$\varepsilon = \frac{\text{COP}_{actual}}{\text{COP}_{rev}} \tag{6.47}$$

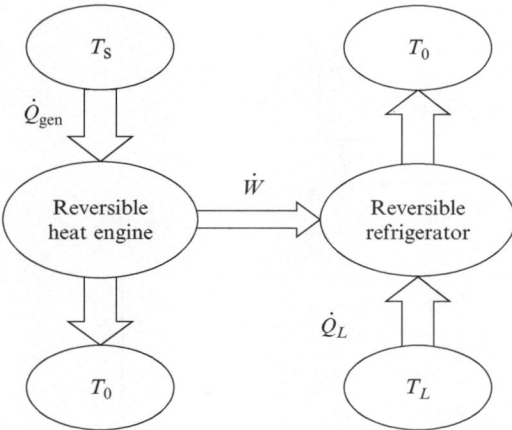

Fig. 6.5 The system used to develop reversible COP of an absorption refrigeration system

6.5 Liquefaction of Gases

Cryogenics is associated with low temperatures, usually defined to be below $-100°C$ (173 K). The general scope of cryogenic engineering is the design, development, and improvement of low-temperature systems and components. The applications of cryogenic engineering include liquefaction of gases, separation of gases, high-field magnets, and sophisticated electronic devices that use the superconductivity property of materials at low temperatures, space simulation, food freezing, medical procedures such as cryogenic surgery, and various chemical processes [36, 37].

The liquefaction of gases has always been an important area of refrigeration inasmuch as many important scientific and engineering processes at cryogenic temperatures depend on liquefied gases. Some examples of such processes are the separation of oxygen and nitrogen from air, preparation of liquid propellants for rockets, study of material properties at low temperatures, and study of some exciting phenomena such as superconductivity. At temperatures above the critical-point value, a substance exists in the gas phase only. The critical temperatures of helium, hydrogen, and nitrogen (three commonly used liquefied gases) are $-268°C$, $-240°C$, and $-147°C$, respectively [1]. Therefore, none of these substances will exist in liquid form at atmospheric conditions. Furthermore, low temperatures of this magnitude cannot be obtained with ordinary refrigeration techniques.

The general principles of various gas liquefaction cycles, including the Linde–Hampson cycle, and their general thermodynamic analyses are presented elsewhere [38–40].

Here we present the methodology for the first and second law based performance analyses of the simple Linde–Hampson cycle, and investigate the effects of gas inlet and liquefaction temperatures on various cycle performance parameters.

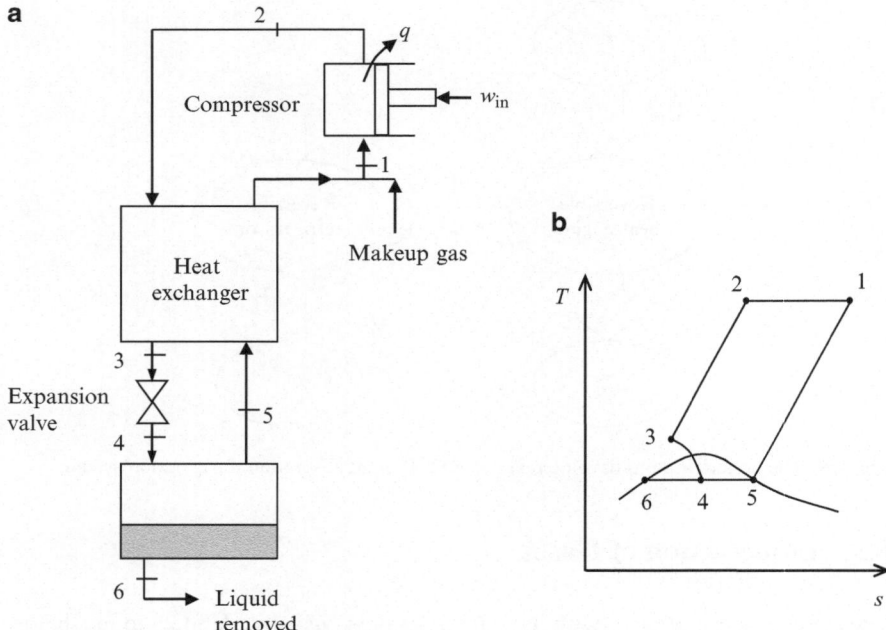

Fig. 6.6 Schematic and temperature-entropy diagram for a simple Linde–Hampson liquefaction cycle [41]

6.5.1 Linde–Hampson Cycle

Several cycles, some complex and others simple, are used successfully for the liquefaction of gases. Here we consider the simple Linde–Hampson cycle, which is shown schematically and on a $T - s$ diagram in Fig. 6.6, in order to describe energy and exergy analyses of liquefaction cycles. See Kanoglu et al. [41] for the details of the analysis in this section. Makeup gas is mixed with the uncondensed portion of the gas from the previous cycle, and the mixture at state 1 is compressed by an isothermal compressor to state 2. The temperature is kept constant by rejecting compression heat to a coolant. The high-pressure gas is further cooled in a regenerative counterflow heat exchanger by the uncondensed portion of gas from the previous cycle to state 3, and is then throttled to state 4, where it is a saturated liquid–vapor mixture. The liquid (state 6) is collected as the desired product, and the vapor (state 5) is routed through the heat exchanger to cool the high-pressure gas approaching the throttling valve. Finally, the gas is mixed with fresh makeup gas, and the cycle is repeated.

The refrigeration effect for this cycle can be defined as the heat removed from the makeup gas in order to turn into a liquid at state 6. Assuming ideal operation for the heat exchanger (i.e., the gas leaving the heat exchanger and the makeup gas are at the same state as state 1, which is the compressor inlet state. This is also the dead-state: $T_1 = T_0$), the refrigeration effect per unit mass of the liquefied gas is given by

$$q_L = h_1 - h_6 = h_1 - h_f \quad \text{(per unit mass of liquefaction)} \quad (6.48)$$

where h_f is the enthalpy of saturated liquid that is withdrawn. From an energy balance on the cycle, the refrigeration effect per unit mass of the gas in the cycle prior to liquefaction may be expressed as

$$q_L = h_1 - h_2 \quad \text{(per unit mass of gas in the cycle)} \quad (6.49)$$

The maximum liquefaction occurs when the difference between h_1 and h_2 (i.e., the refrigeration effect) is maximized. The ratio of (5.2) and (5.1) is the fraction of the gas in the cycle that is liquefied. That is,

$$y = \frac{h_1 - h_2}{h_1 - h_f} \quad (6.50)$$

An energy balance on the heat exchanger gives

$$h_2 - h_3 = x(h_1 - h_5) \quad (6.51)$$

where x is the quality of the mixture at state 4. The fraction of the gas that is liquefied may also be determined from

$$y = 1 - x \quad (6.52)$$

An energy balance on the compressor gives the work of compression per unit mass of the gas in the cycle as

$$w_{\text{actual}} = h_2 - h_1 - T_1(s_2 - s_1) \quad \text{(per unit mass of gas in the cycle)} \quad (6.53)$$

Note that $T_1 = T_0$. The last term in this equation is the isothermal heat rejection from the gas as it is compressed. Considering that the gas generally behaves as an ideal gas during this isothermal compression process, the compression work may also be determined from

$$w_{\text{actual}} = RT_1 \ln(P_2/P_1) \quad (6.54)$$

The coefficient of performance of this cycle is given by

$$\text{COP}_{\text{actual}} = \frac{q_L}{w_{\text{actual}}} = \frac{h_1 - h_2}{h_2 - h_1 - T_1(s_2 - s_1)} \quad (6.55)$$

In liquefaction cycles, one performance parameter used is the work consumed in the cycle for the liquefaction of a unit mass of the gas. This is expressed as

$$w_{actual} = \frac{h_2 - h_1 - T_1(s_2 - s_1)}{y} \text{ (per unit mass of liquefaction)} \tag{6.56}$$

As the liquefaction temperature decreases, the work consumption increases. Noting that different gases have different thermophysical properties and require different liquefaction temperatures, this work parameter should not be used to compare work consumption for the liquefaction of different gases. A reasonable use is to compare different cycles used for the liquefaction of the same gas.

An important object of exergy analysis for systems that consume work such as liquefaction of gases is finding the minimum work required for a certain desired result and comparing it to the actual work consumption. The ratio of these two quantities is often considered the exergy efficiency of such a liquefaction process [42]. Engineers are interested in comparing the actual work used to obtain a unit mass of liquefied gas to the minimum work requirement to obtain the same output. Such a comparison may be performed using the second law of thermodynamics. For instance, the minimum work input requirement (reversible work) and the actual work for a given set of processes may be related to each other by

$$w_{actual} = w_{rev} + T_0 s_{gen} = w_{rev} + x_{dest} \tag{6.57}$$

where T_0 is the environment temperature, s_{gen} is the specific entropy generation, and x_{dest} is the specific exergy destruction during the processes. The reversible work for the simple Linde–Hampson cycle shown in Fig. 6.6 may be expressed by the stream exergy difference of states 1 and 6 as

$$w_{rev} = x_6 - x_1 = h_6 - h_1 - T_0(s_6 - s_1) \tag{6.58}$$

where state 1 has the properties of the makeup gas, which is usually the dead-state. As this equation clearly shows, the minimum work required for liquefaction depends only on the properties of the incoming and outgoing gas being liquefied and the ambient temperature T_0. An exergy efficiency may be defined as the reversible work input divided by the actual work input, both per unit mass of the liquefaction:

$$\varepsilon = \frac{w_{rev}}{w_{actual}} = \frac{h_6 - h_1 - T_0(s_6 - s_1)}{(1/y)[h_2 - h_1 - T_1(s_2 - s_1)]} \tag{6.59}$$

The exergy efficiency may also be defined using actual and reversible COPs of the system as

$$\varepsilon = \frac{COP_{actual}}{COP_{rev}} \tag{6.60}$$

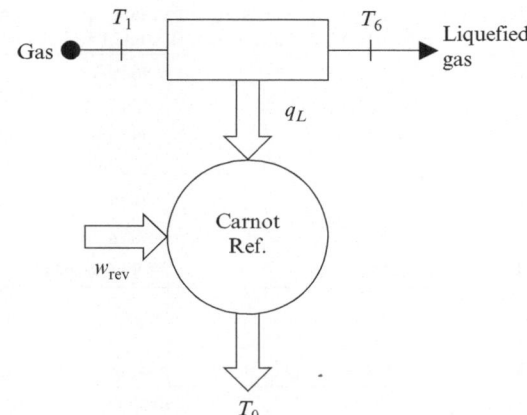

Fig. 6.7 A Carnot refrigerator that uses a minimum amount of work for a liquefaction process

where the reversible COP is given by

$$\text{COP}_{\text{rev}} = \frac{q_L}{w_{\text{rev}}} = \frac{h_1 - h_6}{h_6 - h_1 - T_0(s_6 - s_1)} \tag{6.61}$$

The minimum work input for the liquefaction process is simply the work input required for the operation of a Carnot refrigerator for a given heat removal, which can be expressed as

$$w_{\text{rev}} = \int \delta q \left(1 - \frac{T_0}{T}\right) \tag{6.62}$$

where δq is the differential heat transfer and T is the instantaneous temperature at the boundary where the heat transfer takes place. Note that T is smaller than T_0 for the liquefaction process and to get a positive work input we have to take the sign of heat transfer to be negative because it is a heat output. The evaluation of (6.62) requires knowledge of the functional relationship between the heat transfer δq and the boundary temperature T, which is usually not available. Equation 6.62 is also an expression of the exergy flow associated with the heat removal from the gas being liquefied.

The liquefaction process is essentially the removal of heat from the gas. Therefore, the minimum work can be determined by utilizing a reversible or Carnot refrigerator as shown in Fig. 6.7. The Carnot refrigerator receives heat from the gas and supplies it to the heat sink at T_0 as the gas is cooled from T_1 to T_6. The amount of work that needs to be supplied to this Carnot refrigerator is given by (6.57).

Example 6.4 We present an illustrative example for the simple Linde–Hampson cycle shown in Fig. 6.6. It is assumed that the compressor is reversible and isothermal, the heat exchanger has an effectiveness of 100% (i.e., the gas leaving the liquid reservoir is heated in the heat exchanger to the temperature of the gas

Table 6.1 Various properties and performance parameters of the cycle in Fig. 6.6 for $T_1 = 25°C$, $P_1 = 1$ atm (0.101 MPa), and $P_2 = 20$ MPa. The fluid is air

$h_1 = 298.4$ kJ/kg	$s_1 = 6.86$ kJ/kg · K	$q_L = 424$ kJ/kg liquid
$h_2 = 263.5$ kJ/kg	$s_2 = 5.23$ kJ/kg · K	$w_{actual} = 451$ kJ/kg gas
$h_3 = 61.9$ kJ/kg	$s_f = 2.98$ kJ/kg · K	$w_{actual} = 5481$ kJ/kg liquid
$h_4 = 61.9$ kJ/kg	$T_4 = -194.2°C$	$w_{rev} = 733$ kJ/kg liquid
$h_5 = 78.8$ kJ/kg	$x_4 = 0.9177$	$COP_{acutal} = 0.0775$
$h_6 = -126.1$ kJ/kg	$y = 0.0823$	$COP_{rev} = 0.578$
$h_f = -126.1$ kJ/kg	$q_L = 34.9$ kJ/kg gas	$\varepsilon = 0.134$

Table 6.2 Performance parameters of a simple Linde–Hampson cycle for various fluids

Item	Air	Nitrogen	Oxygen	Argon	Methane	Fluorine
Liquefaction temp. T_4 (°C)	−194.2	−195.8	−183.0	−185.8	−161.5	−188.1
Fraction liquefied y	0.0823	0.0756	0.107	0.122	0.199	0.0765
Refrigeration effect q_L (kJ/kg gas)	34.9	32.6	43.3	33.2	181	26.3
Refrigeration effect q_L (kJ/kg liquid)	424	431	405	272	910	344
Work input w_{in} (kJ/kg gas)	451	468	402	322	773	341
Work input w_{in} (kJ/kg liquid)	5,481	6,193	3,755	2,650	3,889	4,459
Minimum work input w_{rev} (kJ/kg liquid)	733	762	629	472	1080	565
COP_{actual}	0.0775	0.0697	0.108	0.103	0.234	0.0771
COP_{rev}	0.578	0.566	0.644	0.576	0.843	0.609
Exergy efficiency ε, (%)	13.4	12.3	16.8	17.8	27.8	12.7

leaving the compressor), the expansion valve is isenthalpic, and there is no heat leak to the cycle. Furthermore, the gas is taken to be air, at 25°C and 1 atm (0.101 MPa) at the compressor inlet, and the pressure of the gas is 20 MPa at the compressor outlet. With these assumptions and specifications, the various properties at the different states of the cycle and the performance parameters discussed above are determined and listed in Table 6.1. The properties of air and other substances considered are obtained using EES software [20]. This analysis is repeated for different fluids, and the results are listed in Table 6.2.

The COP of a Carnot refrigerator is expressed by the temperatures of the heat reservoirs as

$$COP_{rev} = \frac{1}{T_0/T - 1} \qquad (6.63)$$

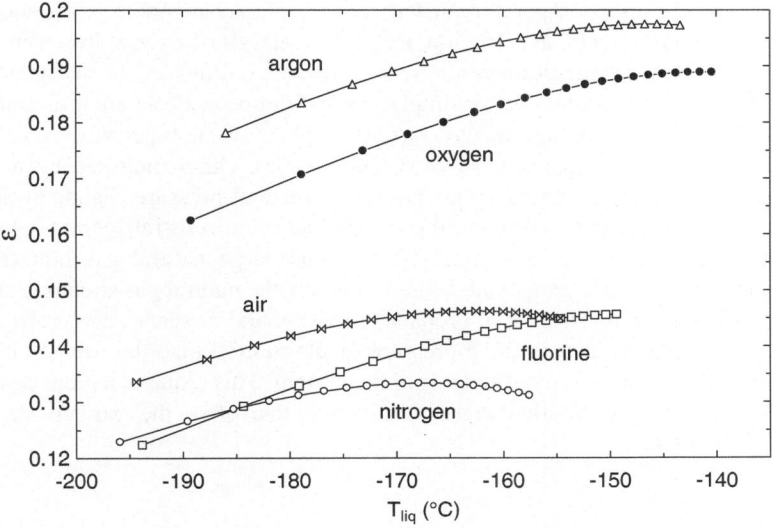

Fig. 6.8 The exergy efficiency versus liquefaction temperature for various gases

Here T represents the temperature of the gas being liquefied in Fig. 6.7, which changes between T_1 and T_6 during the liquefaction process. An average value of T may be obtained using (6.63) with $COP_{rev} = 0.578$ and $T_0 = 25°C$, yielding $T = -156°C$. This is the temperature a heat reservoir would have if a Carnot refrigerator with a COP of 0.578 operated between this reservoir at $-156°C$ and another reservoir at $25°C$. Note that the same reservoir temperature T could be obtained by writing (6.62) in the form

$$w_{rev} = -q_L\left(1 - \frac{T_0}{T}\right) \tag{6.64}$$

where $q_L = 424$ kJ/kg, $w_{rev} = 733$ kJ/kg, and $T_0 = 25°C$.

As part of the analysis, the effects of liquefaction temperature on exergy efficiency are given in Fig. 6.8. The exergy efficiency increases with increasing liquefaction temperature and decreasing inlet gas temperature for all gases considered as shown in Fig. 6.8. In Fig. 6.8, the exergy efficiency reaches a maximum before decreasing at higher temperatures. The decreasing trend at higher liquefaction temperatures is of no practical importance because liquefaction at these high temperatures requires higher inlet pressures, which are not normally used.

The actual exergy efficiencies of the Linde–Hampson liquefaction cycle is usually under 10% [42]. The difference between the actual and reversible work consumption in liquefaction systems is due to exergy losses that occur during various processes in the cycle. Irreversible compression in the compressor, heat transfer across a finite temperature difference in heat exchangers (e.g., regenerator, evaporator, compressor), and friction are major sources of exergy losses in these systems. In actual refrigeration systems, these irreversibilities are normally reduced by applying modifications to the simple Linde–Hampson cycle such as utilizing multistage compression and using a turbine in place of an expansion valve or in conjunction with an expansion valve (Claude cycle). Other modified cycles that have resulted in greater efficiency are known as the dual-pressure Claude cycle and the Collins helium cycle. For natural gas liquefaction, mixed-refrigerant, cascade, and gas-expansion cycles are used [42]. In most large natural gas liquefaction plants, the mixed-refrigerant cycle is used in which the natural gas stream is cooled by the successive vaporization of propane, ethylene, and methane. Each refrigerant may be vaporized at two or three pressure levels to minimize the irreversibilities and thus increase the exergy efficiency of the system. This requires a more complex and costly system but the advantages usually more than offset the extra cost in large liquefaction plants.

6.5.2   Precooled Linde–Hampson Liquefaction Cycle

The precooled Linde–Hampson cycle is a well-known and relatively simple system used for the liquefaction of gases including hydrogen (Fig. 6.9). Makeup gas is mixed with the uncondensed portion of the gas from the previous cycle, and the mixture at state 1 is compressed to state 2. Heat is rejected from the compressed gas to a coolant. The high-pressure gas is cooled to state 3 in a regenerative counterflow

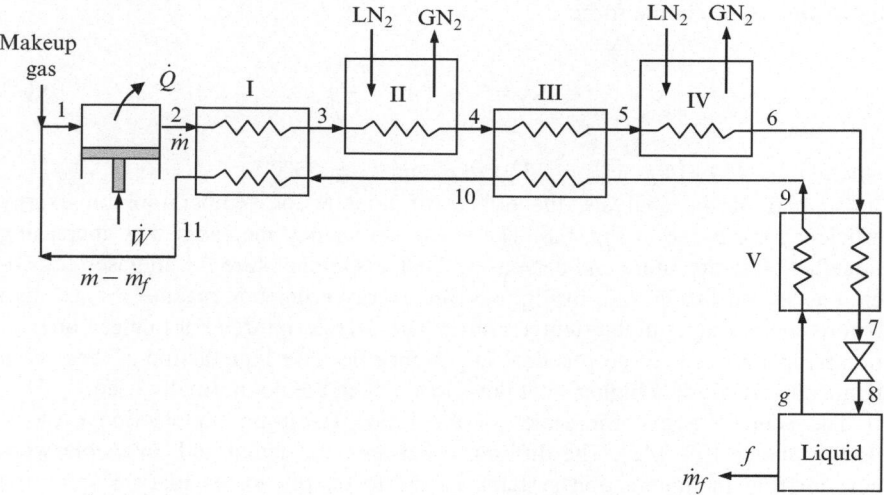

Fig. 6.9 Precooled Linde–Hampson liquefaction cycle

heat exchanger (I) by the uncondensed gas, and is cooled further by flowing through two nitrogen baths (II and IV) and two regenerative heat exchangers (III and V) before being throttled to state 8, where it is a saturated liquid–vapor mixture. The liquid is collected as the desired product, and the vapor is routed through the bottom half of the cycle. Finally, the gas is mixed with fresh makeup gas, and the cycle is repeated.

Using an energy balance of heat exchanger V and the throttling valve taken together, the fraction of the liquefied gas can be determined to be

$$f_{liq} = \frac{h_9 - h_6}{h_9 - h_f} \tag{6.65}$$

Energy balances for the heat exchangers can be written as

$$h_2 - h_3 = (1 - f_{liq})(h_{11} - h_{10}) \tag{6.66}$$

$$h_4 - h_5 = (1 - f_{liq})(h_{10} - h_9) \tag{6.67}$$

$$h_6 - h_7 = (1 - f_{liq})(h_9 - h_g) \tag{6.68}$$

Because the gas behaves ideally during compression, the specific compression work may be determined from

$$w_{in} = \frac{RT_0 \ln(P_2/P_1)}{\eta_{comp}} \text{ (per unit mass of gas in the cycle)} \tag{6.69}$$

where η_{comp} is the isothermal efficiency of the compressor, R is the gas constant, and P is the pressure. The numerator of the right side represents the work input for a corresponding isothermal process. The specific work input to the liquefaction cycle per unit mass of liquefaction is

$$w_{in,liq} = \frac{w_{in}}{f_{liq}} \text{ (per unit mass of liquefaction)} \tag{6.70}$$

A detailed case study on efficiency evaluation of a multistage cascade refrigeration cycle for natural gas liquefaction is given in Appendix B.

Example 6.5 Hydrogen gas at 25°C and 1 atm (101.325 kPa) is to be liquefied in a precooled Linde–Hampson cycle. Hydrogen gas is compressed to a pressure of 10 MPa in the compressor which has an isothermal efficiency of 65%. The effectiveness of heat exchangers is 90%. Determine (a) the heat removed from hydrogen and the minimum work input, (b) the fraction of the gas liquefied, (c) the work input in the compressor per unit mass of liquefied hydrogen, and (d) the second law efficiency of the cycle if the work required for nitrogen liquefaction is

15,200 kJ per kg of hydrogen gas in the cycle. Properties of hydrogen in the cycle at various states are:

$$h_f = 271.1 \, \text{kJ/kg}$$
$$h_0 = 4200 \, \text{kJ/kg}$$
$$h_6 = 965.4 \, \text{kJ/kg}$$
$$h_9 = 1147.7 \, \text{kJ/kg}$$
$$s_f = 17.09 \, \text{kJ/kg} \cdot \text{K}$$
$$s_0 = 70.42 \, \text{kJ/kg} \cdot \text{K}$$

Solution

(a) The heat rejection from hydrogen gas is

$$q_L = h_0 - h_f = (4200 - 271.1) \text{kJ/kg} = \mathbf{3929 \, kJ/kg}$$

Taking the dead-state temperature to be $T_0 = T_1 = 25°C = 298.15$ K, the minimum work input is determined from

$$w_{min} = h_0 - h_f - T_0(s_0 - s_f)$$
$$= (4200 - 271.1) \text{kJ/kg} - (298.15 \text{ K})(70.42 - 17.09) \text{kJ/kg} \cdot \text{K}$$
$$= \mathbf{11,963 \, kJ/kg}$$

(b) The fraction of the gas liquefied is

$$f_{liq} = \frac{h_9 - h_6}{h_9 - h_f} = \frac{1147.7 - 965.4}{1147.7 - 271.1} = \mathbf{0.208}$$

(c) The work input in the compressor per unit mass of hydrogen gas compressed is

$$w_{in} = \frac{RT_0 \ln(P_2/P_1)}{\eta_{comp}} = \frac{(4.124)(298.15) \ln(10,000/101.325)}{0.85} = 8682 \, \text{kJ/kg}$$

Per unit mass of liquefaction,

$$w_{in, liq} = \frac{w_{in}}{f_{liq}} = \frac{8682}{0.208} = \mathbf{41,740 \, kJ/kg}$$

Fig. 6.10 A Carnot
refrigerator operating
between T_L and T_H as
considered in Example 6.6

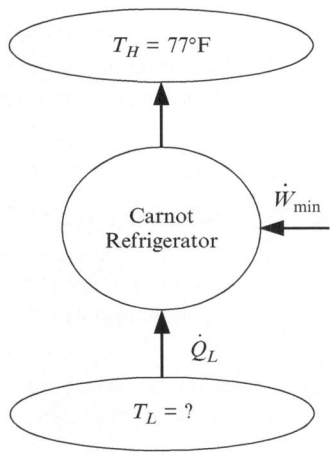

(d) The total work input for the cycle per unit mass of liquefied hydrogen is

$$w_{\text{in, total}} = \frac{w_{\text{in}} + w_{\text{in, nitrogen}}}{f_{\text{liq}}} = \frac{8682 + 15,200}{0.208} = 114,800 \, \text{kJ/kg}$$

(e) The second law efficiency is determined from

$$\varepsilon = \frac{w_{\text{min}}}{w_{\text{in, total}}} = \frac{11,963}{114,800} = 0.104 \text{ or } \mathbf{10.4\%}$$

Example 6.6 Natural gas at 77°F and 1 atm (14.7 psia) at a rate of 2,500 lbm/h is to
be liquefied in a natural gas liquefaction plant. Natural gas leaves the plant at 1 atm
as a saturated liquid. The power consumption is estimated to be 1,650 kW. Using
methane properties for natural gas determine (a) the temperature of natural gas after
the liquefaction process and the rate of heat rejection from the natural gas during
this process, (b) the minimum power input, (c) the actual COP and the second law
efficiency, and (d) the reversible COP. (e) If the liquefaction is done by a Carnot
refrigerator between temperature limits of $T_H = 77°F$ and T_L with the same reversible COP, determine the temperature T_L (see Fig. 6.10). Various properties of
methane before and after the liquefaction process are given as follows.

$$h_1 = -0.4254 \, \text{Btu/lbm}$$
$$h_2 = -391.62 \, \text{Btu/lbm}$$
$$s_1 = -0.0006128 \, \text{Btu/lbm} \cdot \text{R}$$
$$s_2 = -1.5946 \, \text{Btu/lbm} \cdot \text{R}$$

Solution

(a) The state of natural gas after the liquefaction is 14.7 psia and saturated liquid. The temperature at this state is determined from methane tables to be

$$T_2 = -259°\text{F}$$

The rate of heat rejection from the natural gas during the liquefaction process is

$$\dot{Q}_L = \dot{m}(h_1 - h_2) = (2500\,\text{lbm/h})[(-0.4254) - (-391.62)]\text{Btu/lbm}$$
$$= 978,000\,\text{Btu/h}$$

(b) Taking the dead-state temperature to be $T_0 = T_1 = 77°\text{C} = 536\,\text{R}$, the minimum work input is determined from

$$\dot{W}_{\min} = \dot{m}[h_2 - h_1 - T_0(s_2 - s_1)]$$
$$= (2500\,\text{lbm/h})[(-391.62) - (-0.4254)]\text{Btu/lbm} - (537\,\text{R})$$
$$- (-0.0006128)\text{Btu/lbm} \cdot \text{R}]$$
$$= 1.162 \times 10^6\,\text{Btu/h} = 340.5\,\text{kW}$$

(c) The COP and the second law efficiency are

$$\text{COP} = \frac{\dot{Q}_L}{\dot{W}_{\text{actual}}} = \frac{9.78 \times 10^5\,\text{Btu/h}}{(1650\,\text{kW})\left(\frac{3412\,\text{Btu/h}}{1\,\text{kW}}\right)} = 0.1737$$

$$\varepsilon = \frac{\dot{W}_{\min}}{\dot{W}_{\text{actual}}} = \frac{340.5\,\text{kW}}{1650\,\text{kW}} = 0.2064$$

(d) The reversible COP is

$$\text{COP}_{\text{rev}} = \frac{\dot{Q}_L}{\dot{W}_{\min}} = \frac{9.78 \times 10^5\,\text{Btu/h}}{1.162 \times 10^6\,\text{Btu/h}} = 0.842$$

The second law efficiency can also be determined from $\text{COP/COP}_{\text{actual}} = 0.1737/0.842 = 0.206$.

(e) The temperature T_L is determined from

$$\text{COP}_{R,\text{rev}} = \frac{1}{T_H/T_L - 1} \longrightarrow 0.842 = \frac{1}{(537\,\text{R})/T_L - 1} \longrightarrow T_L = 245\,\text{R}$$

It may also be determined from

$$\dot{W}_{\min} = -\dot{Q}_L\left(1 - \frac{T_0}{T_L}\right) \longrightarrow 1.162 \times 10^6\,\text{Btu/h}$$

$$= -(978,000\,\text{Btu/h})\left(1 - \frac{537\,\text{R}}{T_L}\right) \longrightarrow T_L = 245\,\text{R}$$

6.6 Efficiency Analysis of Psychrometric Processes

Psychrometrics is the science of air and water vapor and deals with the properties of moist air. A thorough understanding of psychrometrics is of great significance, particularly to the HVAC&R community. Psychrometrics plays a key role not only in heating and cooling processes and the comfort of the occupants, but in building insulation; roofing properties; and the stability, deformation, and fire-resistance of building materials [37].

In this section, we examine the efficiency aspects of psychrometric processes for HVAC&R and develop mass, energy, entropy, and exergy balances and exergy efficiency relations for some key HVAC&R processes that include heating and cooling, heating with humidification, cooling with dehumidification, evaporative cooling, and adiabatic mixing of air streams. A heating process with humidification is considered as an illustrative example.

6.6.1 Balance Equations for Common Air-Conditioning Processes

In the analysis of air-conditioning processes, four important balances need to be addressed, including mass (i.e., the continuity equation), energy (i.e., the first law of thermodynamics), entropy, and exergy (i.e., the second law of thermodynamics). Air-conditioning processes are essentially steady-flow processes and the general mass, energy, and exergy balances may be written as follows.

Mass balance for dry air:

$$\sum_{\text{in}} \dot{m}_a = \sum_{\text{out}} \dot{m}_a \tag{6.71}$$

Mass balance for water:

$$\sum_{\text{in}} \dot{m}_w = \sum_{\text{out}} \dot{m}_w \quad \text{or} \quad \sum_{\text{in}} \dot{m}_a \omega = \sum_{\text{out}} \dot{m}_a \omega \quad \text{or}$$

$$\dot{m}_w = \dot{m}_a (\omega_{\text{out}} - \omega_{\text{in}}) \tag{6.72}$$

Energy balance (with negligible kinetic and potential energies and work):

$$Q_{\text{in}} + \sum_{\text{in}} \dot{m}h = Q_{\text{out}} + \sum_{\text{out}} \dot{m}h \tag{6.73}$$

Entropy balance (with negligible kinetic and potential energies and work):

$$\dot{S}_{in} - \dot{S}_{out} + \dot{S}_{gen} = 0$$

$$\sum_{in} \dot{S}_{\dot{Q}} + \sum_{in} \dot{m}s - \sum_{out} \dot{S}_{\dot{Q}} - \sum_{out} \dot{m}s + \dot{S}_{gen} = 0 \tag{6.74}$$

$$\sum_{in} \frac{\dot{Q}}{T} + \sum_{in} \dot{m}s - \sum_{out} \frac{\dot{Q}}{T} - \sum_{out} \dot{m}s + \dot{S}_{gen} = 0$$

Exergy balance (with negligible kinetic and potential energies and work):

$$\sum_{in} \dot{X}_{\dot{Q}} + \sum_{in} \dot{m}\psi - \sum_{out} \dot{X}_{\dot{Q}} - \sum_{out} \dot{m}\psi - \dot{X}_{dest} = 0$$

$$\sum_{in} \dot{Q}\left(1 - \frac{T_0}{T}\right) + \sum_{in} \dot{m}\psi - \sum_{out} \dot{Q}\left(1 - \frac{T_0}{T}\right) - \sum_{out} \dot{m}\psi - \dot{X}_{dest} = 0 \tag{6.75}$$

The stream flow exergy is given by

$$\psi = h - h_0 - T_0(s - s_0) \tag{6.76}$$

The exergy destruction is proportional to entropy generation due to irreversibilities, and can be expressed as

$$\dot{X}_{dest} = T_0 \dot{S}_{gen} \tag{6.77}$$

Equation 6.76 is useful when the enthalpy and entropy of moist air can be obtained directly from a property database. Alternatively, the stream flow exergy can be determined by considering dry air and water vapor as an ideal gas to be [43]

$$\psi = (c_{p,a} + \omega c_{p,v})T_0\left(\frac{T}{T_0} - 1 - \ln\frac{T}{T_0}\right) + (1 + \tilde{\omega})R_a T_0 \ln\frac{P}{P_0}$$

$$+ R_a T_0\left[(1 + \tilde{\omega})\ln\frac{1 + \tilde{\omega}_o}{1 + \tilde{\omega}} + \tilde{\omega}\ln\frac{\tilde{\omega}}{\tilde{\omega}_o}\right] \tag{6.78}$$

where the last term is the specific chemical exergy. The proportionality between specific humidity ratio ω and specific humidity ratio on a molar basis $\tilde{\omega}$ is given by

$$\tilde{\omega} = 1.608\omega \tag{6.79}$$

where the humidity ratio is

$$\omega = m_v/m_a \tag{6.80}$$

Fig. 6.11 Simple heating or cooling process as represented on a psychrometric chart

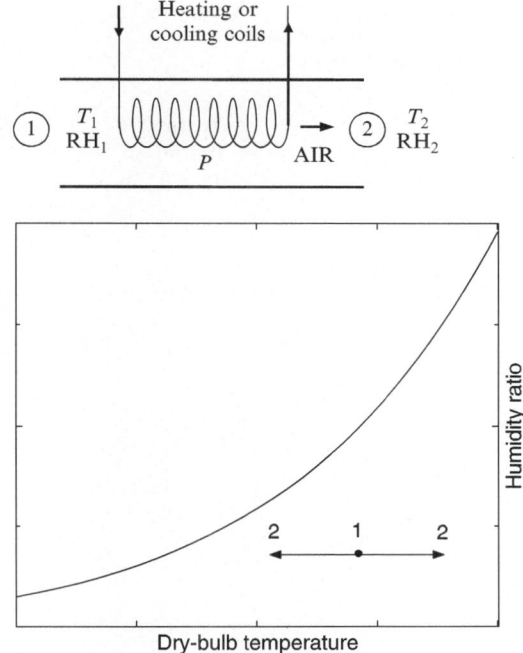

The exergy efficiency of an air-conditioning process may be written as

$$\varepsilon = \frac{\dot{X}_{\text{out}}}{\dot{X}_{\text{in}}} = 1 - \frac{\dot{X}_{\text{dest}}}{\dot{X}_{\text{in}}} \tag{6.81}$$

where $\dot{X}_{\text{out}}$ and $\dot{X}_{\text{in}}$ are the exergy output and input for the component, respectively, and $\dot{X}_{\text{dest}}$ is the exergy destruction during the process.

Common air-conditioning processes are shown schematically and on a psychrometric chart in Figs. 6.11 through 6.15. Cengel and Boles [1] provide mass and energy balances of these processes. They are also extensively discussed in Kanoglu et al. [44]. Figure 6.11 shows a heating or cooling process during which only a change in sensible heat is encountered. There is no change in latent heat due to the constant humidity ratio of the air. Figure 6.12 is an example of a heating and humidification process. Air is first heated in a heating section (process 1-2) and then humidified (process 2-3) by the injection of steam. In the cooling and dehumidification process shown in Fig. 6.13, air is cooled at a constant humidity ratio until it is saturated (process 1-x). Further cooling of air (process x-2) results in dehumidification. Figure 6.14 illustrates adiabatic humidification (i.e., evaporative cooling) at a constant wet-bulb temperature, as occurs in spray type humidification. In this process, water should be injected at the temperature of the exiting air. Figure 6.15 shows a mixing process of two streams of air (one at state 1 and other at 2, and the resulting mixture reaches state 3).

Fig. 6.12 Heating with humidification as represented on a psychrometric chart

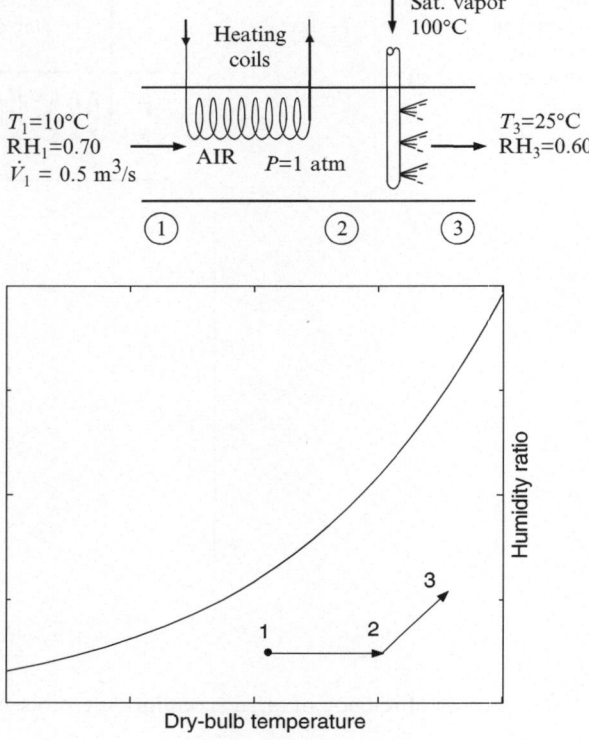

Balances for each of these air-conditioning processes are developed, based on the general formulations given in (6.70) through (6.81). Subscripts refer to the associated state points in Figs. 6.11 through 6.15 and the state of water is represented by the subscript w.

6.6.2 Heating or Cooling

$$\text{Dry air mass balance:} \quad \dot{m}_{a1} = \dot{m}_{a2} \tag{6.82}$$

$$\text{Water mass balance:} \quad \dot{m}_{w1} = \dot{m}_{w2} \tag{6.83}$$

$$\text{Energy balance:} \quad \dot{Q}_{\text{in}} + \dot{m}_{a1}h_1 = \dot{m}_{a2}h_2 \quad \text{(heating)} \tag{6.84}$$

$$\dot{m}_{a1}h_1 = \dot{m}_{a2}h_2 + \dot{Q}_{\text{out}} \quad \text{(cooling)} \tag{6.85}$$

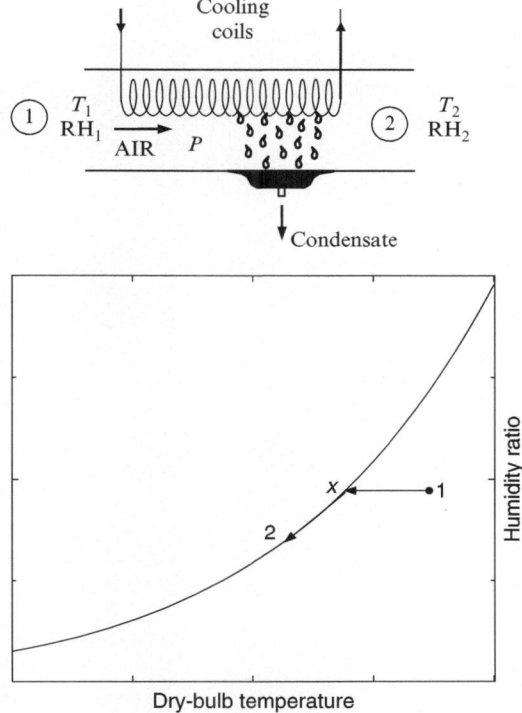

Fig. 6.13 Cooling with dehumidification as represented on a psychrometric chart

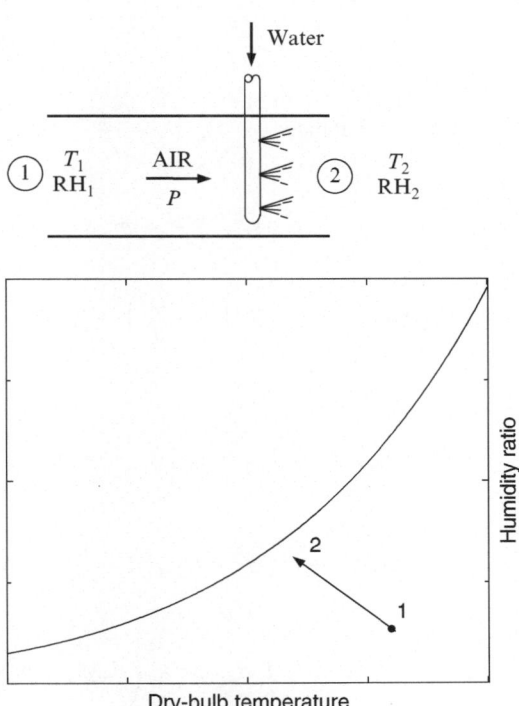

Fig. 6.14 Evaporative cooling as represented on a psychrometric chart

Fig. 6.15 Adiabatic mixing
of airstreams as represented
on a psychrometric chart

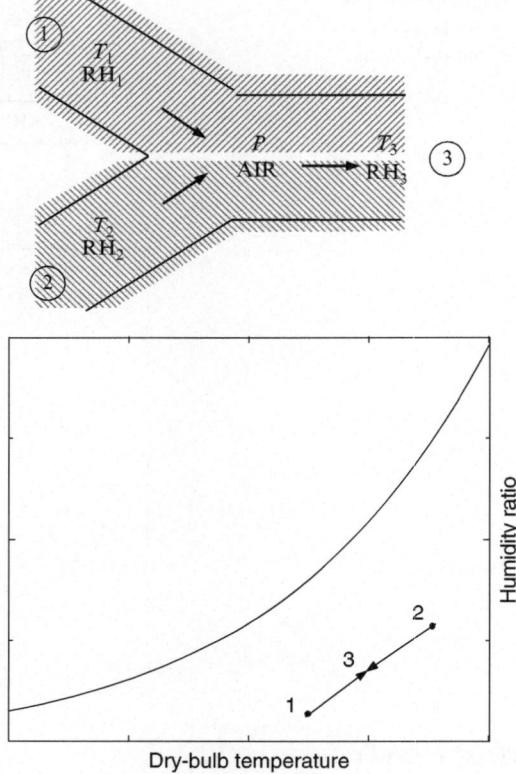

Entropy balance: $\qquad \dot{m}_{a1}s_1 + \dfrac{\dot{Q}_{\text{in}}}{T} - \dot{m}_{a2}s_2 + \dot{S}_{\text{gen}} = 0$ (heating) $\qquad$ (6.86)

$$\dot{m}_{a1}s_1 - \dfrac{\dot{Q}_{\text{out}}}{T} - \dot{m}_{a2}s_2 + \dot{S}_{\text{gen}} = 0 \quad \text{(cooling)} \qquad (6.87)$$

Exergy balance: $\qquad \dot{Q}_{\text{in}}\left(1 - \dfrac{T_0}{T}\right) + \dot{m}_{a1}\psi_1 - \dot{m}_{a2}\psi_2 - \dot{E}x_{\text{dest}} = 0$ (heating)

$$(6.88)$$

$$\dot{X}_{\text{dest}} = T_0\dot{S}_{\text{gen}} = T_0\left(\dot{m}_{a2}s_2 - \dot{m}_{a1}s_1 - \dfrac{\dot{Q}_{\text{in}}}{T}\right) \quad \text{(heating)} \qquad (6.89)$$

$$\dot{m}_{a1}\psi_1 - \dot{m}_{a2}\psi_2 - \dot{Q}_{\text{out}}\left(1 - \dfrac{T_0}{T}\right) - \dot{X}_{\text{dest}} = 0 \quad \text{(cooling)} \qquad (6.90)$$

$$\dot{X}_{\text{dest}} = T_0\dot{S}_{\text{gen}} = T_0\left(\dot{m}_{a2}s_2 - \dot{m}_{a1}s_1 - \dfrac{\dot{Q}_{\text{out}}}{T}\right) \text{(cooling)} \qquad (6.91)$$

$$\text{Exergy efficiency:}\quad \varepsilon_{\text{heating}} = \frac{\dot{m}_{a2}\psi_2}{\dot{m}_{a1}\psi_1 + \dot{Q}_{\text{in}}\left(1 - \dfrac{T_0}{T}\right)} \qquad (6.92)$$

$$\varepsilon_{\text{cooling}} = \frac{\dot{m}_{a2}\psi_2 + \dot{Q}_{\text{out}}\left(1 - \dfrac{T_0}{T}\right)}{\dot{m}_{a1}\psi_1} \qquad (6.93)$$

6.6.3   Heating with Humidification

$$\text{Dry air mass balance: } \dot{m}_{a1} = \dot{m}_{a2} = \dot{m}_{a3} \qquad (6.94)$$

$$\text{Water mass balance: } \dot{m}_{w1} = \dot{m}_{w2} \qquad (6.95)$$

$$\dot{m}_{w2} + \dot{m}_w = \dot{m}_{w3} \longrightarrow \dot{m}_{a2}\omega_2 + \dot{m}_w = \dot{m}_{a3}\omega_3 \qquad (6.96)$$

$$\text{Energy balance:}\quad \dot{Q}_{\text{in}} + \dot{m}_{a1}h_1 = \dot{m}_{a2}h_2 \quad (\text{process } 1\text{--}2) \qquad (6.97)$$

$$\dot{m}_{a2}h_2 + \dot{m}_w h_w = \dot{m}_{a3}h_3 \quad (\text{process } 2\text{--}3) \qquad (6.98)$$

$$\dot{Q}_{\text{in}} + \dot{m}_{a1}h_1 + \dot{m}_w h_w = \dot{m}_{a3}h_3 \quad (\text{process } 1\text{--}3) \qquad (6.99)$$

$$\text{Entropy balance:}\quad \dot{m}_{a1}s_1 + \dot{m}_w s_w + \frac{\dot{Q}_{\text{in}}}{T} - \dot{m}_{a3}s_3 + \dot{S}_{\text{gen}} = 0 \quad (\text{process } 1\text{--}3) \qquad (6.100)$$

$$\text{Exergy balance:}\quad \dot{Q}_{\text{in}}\left(1 - \frac{T_0}{T}\right) + \dot{m}_{a1}\psi_1 - \dot{m}_{a2}\psi_2 - \dot{X}_{\text{dest}} = 0 \quad (\text{process } 1\text{--}2) \qquad (6.101)$$

$$\dot{m}_{a2}\psi_2 + \dot{m}_w \psi_w - \dot{m}_{a3}\psi_3 - \dot{X}_{\text{dest}} = 0 \quad (\text{process } 2\text{--}3) \qquad (6.102)$$

$$\dot{Q}_{\text{in}}\left(1 - \frac{T_0}{T}\right) + \dot{m}_{a1}\psi_1 + \dot{m}_w \psi_w - \dot{m}_{a3}\psi_3 - \dot{X}_{\text{dest}} = 0 \quad (\text{process } 1\text{--}3) \quad (6.103)$$

$$\dot{X}_{\text{dest}} = T_0\dot{S}_{\text{gen}} = T_0\left(\dot{m}_{a3}s_3 - \dot{m}_{a1}s_1 - \dot{m}_w s_w - \frac{\dot{Q}_{\text{in}}}{T}\right) \quad (\text{process } 1\text{--}3) \quad (6.104)$$

$$\text{Exergy efficiency:}\quad \varepsilon = \frac{\dot{m}_{a3}\psi_3}{\dot{Q}_{\text{in}}\left(1 - \dfrac{T_0}{T}\right) + \dot{m}_{a1}\psi_1 + \dot{m}_w \psi_w} \qquad (6.105)$$

6.6.4 Cooling with Dehumidification

$$\text{Dry air mass balance:} \quad \dot{m}_{a1} = \dot{m}_{a2} \qquad (6.106)$$

$$\text{Water mass balance:} \quad \dot{m}_{w1} = \dot{m}_{w2} + \dot{m}_w \longrightarrow \dot{m}_{a1}\omega_1 = \dot{m}_{a2}\omega_2 + \dot{m}_w \qquad (6.107)$$

$$\text{Energy balance:} \quad \dot{m}_{a1}h_1 = \dot{Q}_{\text{out}} + \dot{m}_{a2}h_2 + \dot{m}_w h_w \qquad (6.108)$$

$$\text{Entropy balance:} \quad \dot{m}_{a1}s_1 - \dot{m}_w s_w - \frac{\dot{Q}_{\text{out}}}{T} - \dot{m}_{a2}s_2 + \dot{S}_{\text{gen}} = 0 \qquad (6.109)$$

Exergy balance:

$$\dot{m}_{a1}\psi_1 - \dot{Q}_{\text{out}}\left(1 - \frac{T_0}{T}\right) - \dot{m}_{a2}\psi_2 - \dot{m}_w\psi_w - \dot{X}_{\text{dest}} = 0 \qquad (6.110)$$

$$\dot{X}_{\text{dest}} = T_0\dot{S}_{\text{gen}} = T_0\left(\dot{m}_{a2}s_2 + \dot{m}_w s_w + \frac{\dot{Q}_{\text{out}}}{T} - \dot{m}_{a1}s_1\right) \qquad (6.111)$$

$$\text{Exergy efficiency:} \quad \varepsilon \frac{\dot{Q}_{\text{out}}\left(1 - \frac{T_0}{T}\right) + \dot{m}_{a2}\psi_2 + \dot{m}_w\psi_w}{\dot{m}_{a1}\psi_1} \qquad (6.112)$$

6.6.5 Evaporative Cooling

$$\text{Dry air mass balance:} \quad \dot{m}_{a1} = \dot{m}_{a2} \qquad (6.113)$$

$$\text{Water mass balance:} \quad \dot{m}_{w1} + \dot{m}_w = \dot{m}_{w2} \longrightarrow \dot{m}_{a1}\omega_1 + \dot{m}_w = \dot{m}_{a2}\omega_2 \qquad (6.114)$$

$$\text{Energy balance:} \quad \dot{m}_{a1}h_1 = \dot{m}_{a2}h_2 \longrightarrow h_1 = h_2 \qquad (6.115)$$

$$\dot{m}_{a1}h_1 + \dot{m}_w h_w = \dot{m}_{a2}h_2 \qquad (6.116)$$

$$\text{Entropy balance:} \quad \dot{m}_{a1}s_1 + \dot{m}_w s_w - \dot{m}_{a2}s_2 + \dot{S}_{\text{gen}} = 0 \qquad (6.117)$$

$$\text{Exergy balance:} \quad \dot{m}_{a1}\psi_1 + \dot{m}_w\psi_w - \dot{m}_{a2}\psi_2 - \dot{X}_{\text{dest}} = 0 \qquad (6.118)$$

$$\dot{X}_{\text{dest}} = T_0\dot{S}_{\text{gen}} = T_0(\dot{m}_{a2}s_2 - \dot{m}_{a1}s_1 - \dot{m}_w s_w) \qquad (6.119)$$

$$\text{Exergy efficiency:} \quad \varepsilon = \frac{\dot{m}_{a2}\psi_2}{\dot{m}_{a1}\psi_1 + \dot{m}_w\psi_w} \qquad (6.120)$$

6.6.6 Adiabatic Mixing of Air Streams

$$\text{Dry air mass balance: } \dot{m}_{a1} + \dot{m}_{a2} = \dot{m}_{a3} \tag{6.121}$$

$$\text{Water mass balance: } \dot{m}_{w1} + \dot{m}_{w2} = \dot{m}_{w3} \longrightarrow \dot{m}_{a1}\omega_1 + \dot{m}_{a2}\omega_2 = \dot{m}_{a3}\omega_3 \tag{6.122}$$

$$\text{Energy balance: } \dot{m}_{a1}h_1 + \dot{m}_{a2}h_2 = \dot{m}_{a3}h_3 \tag{6.123}$$

$$\text{Entropy balance: } \dot{m}_{a1}s_1 + \dot{m}_{a2}s_2 - \dot{m}_{a3}s_3 + \dot{S}_{gen} = 0 \tag{6.124}$$

$$\text{Exergy balance: } \dot{m}_{a1}\psi_1 + \dot{m}_{a2}\psi_2 - \dot{m}_{a3}\psi_3 - \dot{X}_{dest} = 0 \tag{6.125}$$

$$\dot{X}_{dest} = T_0\dot{S}_{gen} = T_0(\dot{m}_{a3}s_3 - \dot{m}_{a1}s_1 - \dot{m}_{a2}s_2) \tag{6.126}$$

$$\text{Exergy efficiency: } \varepsilon = \frac{\dot{m}_{a3}\psi_3}{\dot{m}_{a1}\psi_1 + \dot{m}_{a2}\psi_2} \tag{6.127}$$

Example 6.7 A heating process with humidification is considered using the values shown in Fig. 6.12. Based on the balances [(6.94) through (6.105)] and using an equation solver with built-in thermodynamic functions [20], we obtain following results.

$$\dot{m}_a = 0.618\,\text{kg/s}, \; \dot{m}_w = 0.00406\,\text{kg/s}, \; T_2 = 24.2°\text{C}, \; \dot{Q}_{in} = 8.90\,\text{kW}$$

$$\dot{X}_{in} = 4.565\,\text{kW}, \; \dot{X}_{dest} = 4.238\,\text{kW}, \; \varepsilon = 0.0718 \text{ or } 7.2\%$$

The dead-state properties of air are taken to be the same as the inlet air properties whereas the dead-state properties of water are obtained using the temperature of inlet air and the atmospheric pressure. The temperature at which heat transfer takes place is assumed to be equal to the temperature of the saturated water vapor used for humidification. When property data for the fluid flowing in the heating coil are available, we do not have to assume a temperature for heat transfer. For example, let us assume that a refrigerant flows in the heating coil and the properties of the refrigerant at the inlet (denoted by subscript $R1$) and exit (denoted by subscript $R2$) of the heating section are given. The balances in this case become

$$\text{Energy balance: } \dot{Q}_{in} + \dot{m}_{a1}h_1 = \dot{m}_{a2}h_2 \text{ (process 1–2)}$$

$$\dot{m}_{a1}h_1 + \dot{m}_R h_{R1} = \dot{m}_{a2}h_2 + \dot{m}_R h_{R2} \text{ (process 1–2)}$$

$$\dot{m}_{a1}h_1 + \dot{m}_R h_{R1} + \dot{m}_w h_w = \dot{m}_{a2}h_2 + \dot{m}_R h_{R2} \text{ (process 1–3)}$$

Entropy balance:

$$\dot{m}_{a1}s_1 + \dot{m}_R s_{R1} + \dot{m}_w s_w - \dot{m}_{a3}s_3 - \dot{m}_R s_{R2} + \dot{S}_{gen} = 0 \text{ (process } 1-3)$$

Exergy balance:

$$\dot{m}_{a1}\psi_1 + \dot{m}_R \psi_{R1} + \dot{m}_w \psi_w - \dot{m}_{a3}\psi_3 - \dot{m}_R \psi_{R2} - \dot{X}_{dest} = 0 \text{ (process } 1-3)$$

$$\dot{X}_{dest} = T_0 \dot{S}_{gen} = T_0(\dot{m}_{a3}s_3 + \dot{m}_R s_{R2} - \dot{m}_{a1}s_1 - \dot{m}_R s_{R1} - \dot{m}_w s_w)(\text{process } 1-3)$$

$$\text{Exergy efficiency: } \varepsilon = \frac{\dot{m}_{a3}\psi_3 + \dot{m}_R \psi_{R2}}{\dot{m}_{a1}\psi_1 + \dot{m}_R \psi_{R1} + \dot{m}_w \psi_w}$$

The exergy efficiency of the process is calculated to be 7.2%, which is low. This is typical of air-conditioning processes during which irreversibilities occur mainly due to heat transfer across a relatively high temperature difference and humidification. The effect of some of the operating parameters on system performance is investigated in Kanoglu et al. [44].

Appendix A

This case study is based on the study by

M. Kanoğlu and A. Bolatturk: Performance and parametric investigation of a binary geothermal power plant by exergy, in *Renewable Energy*, 33: 2366–2374, 2008.

A.1 A Case Study of Efficiency Evaluation of a Binary Geothermal Power Plant

A.1.1 Introduction

Geothermal energy is widely used as a reliable source of electricity generation. Geothermal plants are in operation in 21 countries and have a combined installed capacity of over 6,000 MW. Electricity has been generated from geothermal resources since the early 1960s. Most of the world's geothermal power plants were built in the 1970s and 1980s following the 1973 oil crisis. The urgency to generate electricity from alternative energy sources and the fact that geothermal energy was essentially free led to nonoptimal plant designs for using geothermal resources.

Three major types of power plants are operating today: dry-steam plants, flash-steam plants, and binary-cycle plants where binary and combined flash/binary plants are relatively new designs. Even though new geothermal power plants are being built using current technologies such as combined flash/binary cycles, not many new geothermal power plants are expected to be built. The thermal efficiencies of conventional combustion-based power plants have increased significantly in recent decades with the use of combined cycles. The initial cost of building a geothermal power plant has increased over the years. There is, however, a great potential to increase efficiencies of some existing binary geothermal power plants by replacing the binary fluid for a better match between the changing resource conditions and the power generation equipment, using the moderate-temperature reinjected brine for heating and cooling applications, and considering cogeneration and other means when possible. Also, supercritical cycles for geo-thermal power generation systems were studied to raise the power output and thermal efficiency by selecting the most suitable working fluids and optimizing the cyclic parameters.

Geothermal energy is used to generate electricity and for direct uses such as space heating and cooling, industrial processes, and greenhouse heating. High-temperature geothermal resources above 150°C are generally used for power generation. Moderate-temperature (between 90°C and 150°C) and low-temperature (below 90°C) geothermal resources are best suited for direct uses. However, geothermal energy is more effective when used directly than when converted to electricity, particularly for moderate- and low-temperature geothermal resources because the direct use of geothermal heat for heating and cooling would replace the burning of fossil fuels from which electricity is generated much more efficiently.

Exergy analysis based on the second law of thermodynamics has proven to be a powerful tool in performance evaluation and the thermodynamic analysis of energy systems. This also applies to performance evaluation of geothermal power plants. The temperatures of geothermal fluids are relatively low, so the first law efficiencies of geothermal power plants are also inherently low. Consequently, the difference between the first law efficiency of a good performing and that of a poorly performing geothermal power plant located at similar sites is small. It then becomes difficult to make a comparison on the basis of first law efficiencies only. This is especially true in binary geothermal power plants inasmuch as the resource tem-perature is lower than in single-and double-flash systems.

In this case study, a geothermal power plant in Reno, Nevada, United States, is considered and the exergy analysis of the plant is performed. The exergy and energy efficiencies are calculated for both the entire plant and for the individual plant components. The sites of exergy destruction are identified and quantified. Also, the effects of turbine inlet pressure and temperature and the condenser pressure on exergy and energy efficiencies, the net power output, and the brine reinjection temperature are investigated and the trends are explained.

A.1.2   Plant Operation

Binary cycle plants use the geothermal brine from liquid-dominated resources. These plants operate with a binary working fluid (isobutane, isopentane, R-114, etc.) that has a low boiling temperature in a Rankine cycle. The working fluid is completely vaporized and usually superheated by the geothermal heat in the vaporizer. The vapor expands in the turbine, and is then condensed in an air- or water-cooled condenser before being pumped back to the vaporizer to complete the cycle.

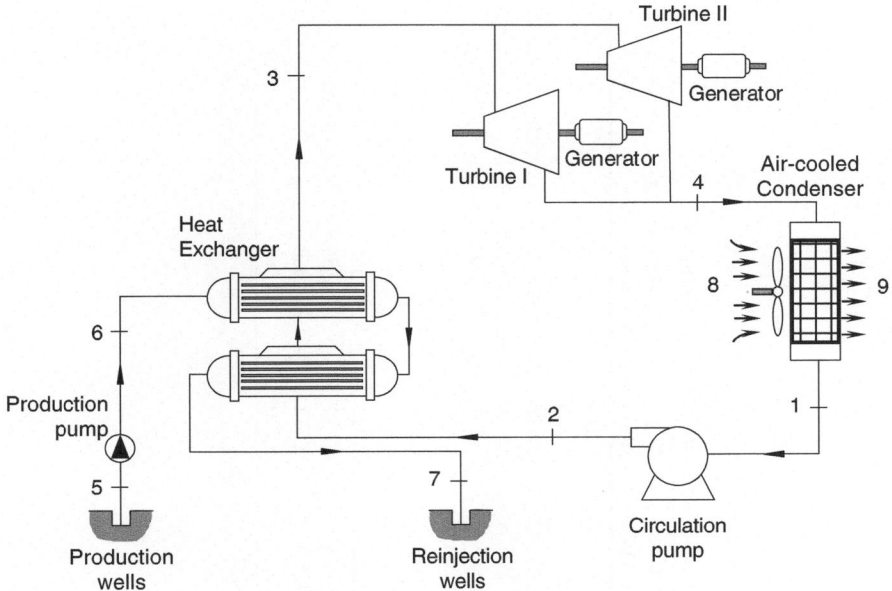

Fig. A.1 Schematic layout of the plant

The geothermal power plant analyzed is a binary design plant that generates a yearly average net power output of about 27 MW. The plant consists of two identical units, each having two identical turbines. Only one unit is considered for the rest of the appendix. The schematic of one unit and the properties at various states are given in Fig. A.1 and Table A.1, respectively. Brine is extracted from five production wells whose average depth is about 160 m. The power plant operates on a liquid-dominated resource at 160°C. The brine passes through the heat exchanger system that consists of a series of counterflow heat exchangers where heat is transferred to the working (binary) fluid isobutane before the brine is reinjected back to the ground. Isobutane is found superheated at the heat exchanger exit. An equal amount of isobutane flows through each turbine. The mechanical power extracted from the turbines is converted to electrical power in generators. It utilizes a dry-air condenser to condense the working fluid so no fresh water is consumed. Isobutane circulates in a closed cycle, which is based on the Rankine cycle.

The harvested geothermal fluid is saturated liquid at 160°C and 1,264 kPa in the reservoir. The heat source for the plant is the flow of geothermal water (brine) entering the plant at 158°C and 609 kPa with a total mass flow rate of 555.9 kg/s. Geothermal fluid remains as a liquid throughout the plant. The brine leaving the heat exchangers is directed to the reinjection wells where it is reinjected back into the ground at 90°C and 423 kPa.

Table A.1 Exergy rates and other properties at various plant locations for one representative unit. State numbers refer to Fig. A.1

State no	Fluid	Phase	Temperature T (°C)	Pressure P (kPa)	Enthalpy h (kJ/kg)	Entropy s (kJ/kg°C)	Mass flow rate $\dot{m}$ (kg/s)	Exergy rate $\dot{E}$ (kW)
0	Brine	Dead-state	3.0	84	12.6	0.046	–	–
0'	Isobutane	Dead-state	3.0	84	207	1.025	–	–
0''	Air	Dead-state	3.0	84	276.5	5.672	–	–
1	Isobutane	Comp. liquid	11.7	410	227.5	1.097	305.6	221.3
2	Isobutane	Comp. liquid	13.76	3,250	234.3	1.103	305.6	1,768
3	Isobutane	Sup. vapor	146.8	3,250	760.9	2.544	305.6	41,094
4	Isobutane	Sup. vapor	79.5	410	689.7	2.601	305.6	14,528
5	Brine	Liquid	160.0	1,264	675.9	1.942	555.9	77,656
6	Brine	Liquid	158.0	609	666.8	1.923	555.9	75,586
7	Brine	Liquid	90.0	423	377.3	1.193	555.9	26,706
8	Air	Gas	3.0	84	276.5	5.672	8,580	0
9	Air	Gas	19.4	84	292.9	5.730	8,580	4,036

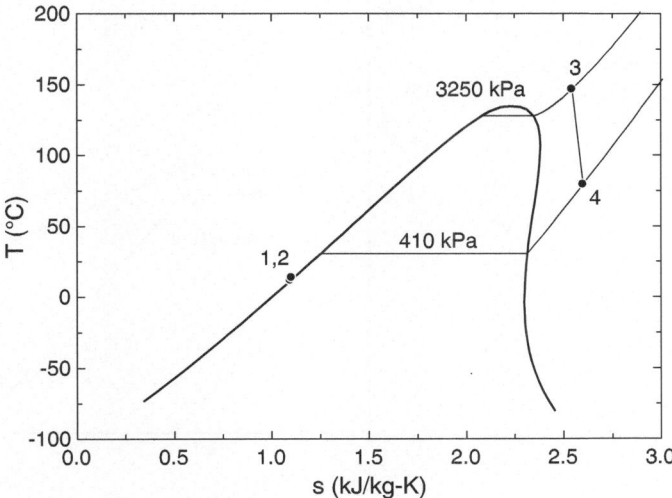

Fig. A.2 Temperature–entropy (*T*–*s*) diagram of binary Rankine cycle

In the plant, the mass flow rate 305.6 kg/s of working fluid circulates through the cycle. The working fluid enters the heat exchanger at 13.7°C and leaves after it is evaporated at 128°C and superheated to 146.8°C. The working fluid then passes through the turbines at each mass flow rate of 152.8 kg/s. It exhausts to an air-cooled condenser at about 79.5°C where it condenses to a temperature of 11.7°C. Approximately 8,580 kg/s air at an ambient temperature of 3°C is required to absorb the heat yielded by the working fluid. This raises the air temperature to 19.4°C. The working fluid is pumped to the heat exchanger pressure to complete the Rankine cycle. The isobutane cycle on a *T* – *s* diagram is shown in Fig. A.2. It is noted in Fig. A.2 that the saturated vapor line of isobutane has a positive slope ensuring a superheated vapor state at the turbine outlet. Thus, no moisture is involved in the turbine operation. This is one reason isobutane is a suitable working fluid in binary geothermal power plants.

The heat exchange process between the geothermal brine and working fluid isobutane is shown in Fig. A.3. An energy balance can be written from Fig. A.3 for the heat exchanger as

$$\dot{m}_{geo}(h_6 - h_{pp}) = \dot{m}_{binary}(h_3 - h_{f,binary}) \tag{A.1}$$

and

$$\dot{m}_{geo}(h_{pp} - h_7) = \dot{m}_{binary}(h_{f,binary} - h_2) \tag{A.2}$$

Fig. A.3 Diagram showing the heat exchange process between the geothermal brine and the working fluid isobutane in the heat exchanger

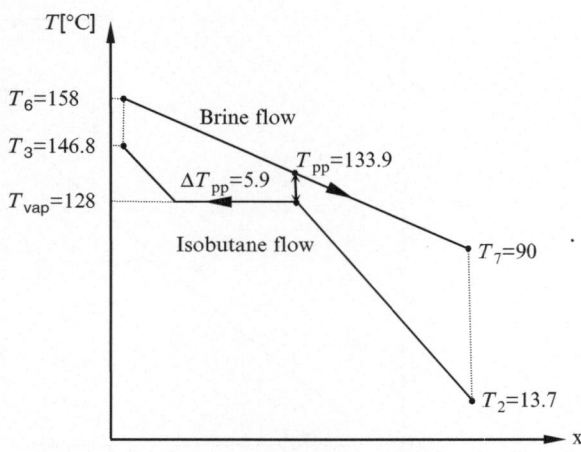

where $\dot{m}_{\text{geo}}$ and $\dot{m}_{\text{binary}}$ are the mass flow rate of geothermal brine and binary fluid, respectively. Also, $h_{\text{f,binary}}$ is the saturated liquid enthalpy of isobutane at the saturated (vaporization) temperature, T_{vap} to be 128°C, and h_{pp} is the enthalpy of the brine at the pinch-point temperature of the brine. Solving these equations for h_{pp}, we determine the corresponding brine pinch-point temperature T_{pp} to be 133.9°C. The pinch-point temperature difference ΔT_{pp} is simply the difference between the brine pinch-point temperature and the vaporization temperature of isobutane, resulting in 5.9°C.

A.1.3 Energy and Exergy Analyses

Mass, energy, and exergy balances for any control volume at steady-state with negligible kinetic and potential energy changes can be expressed, respectively, by

$$\sum \dot{m}_{\text{in}} = \sum \dot{m}_{\text{out}} \tag{A.3}$$

$$\dot{Q} + \dot{W} = \sum \dot{m}_{\text{out}} h_{\text{out}} - \sum \dot{m}_{\text{in}} h_{\text{in}} \tag{A.4}$$

$$\dot{E}_{\text{heat}} + \dot{W} = \sum \dot{E}_{\text{out}} - \sum \dot{E}_{\text{in}} + \dot{I} \tag{A.5}$$

where the subscripts in and out represent the inlet and exit states, $\dot{Q}$ and $\dot{W}$ are the net heat and work inputs, $\dot{m}$ is the mass flow rate, h is the enthalpy, and $\dot{I}$ is the rate of irreversibility (exergy destruction). The subscript 0 stands for the restricted dead-state. Also, $\dot{E}_{\text{heat}}$ is the net exergy transfer by heat at the temperature T, which is given by

$$\dot{E}_{\text{heat}} = \sum \left(1 - \frac{T_0}{T}\right)\dot{Q} \qquad (A.6)$$

The specific flow exergy is given by

$$e = h - h_0 - T_0(s - s_0) \qquad (A.7)$$

Multiplying specific exergy by the mass flow rate of the fluid gives the exergy rate

$$\dot{E} = \dot{m}e \qquad (A.8)$$

The exergetic efficiency of a turbine is defined as a measure of how well the stream exergy of the fluid is converted into actual turbine output. Then,

$$\eta_{\text{ex,turb}} = \frac{\dot{W}_{\text{turb}}}{\dot{W}_{\text{turb,rev}}} \qquad (A.9)$$

where $\dot{W}_{\text{turb}}$ is the actual turbine power and $\dot{W}_{\text{turb,rev}}$ is the reversible turbine power, which is equal to $\dot{W}_{\text{turb}} + \dot{I}$. The exergy efficiency of the compressor is defined similarly as

$$\eta_{\text{ex,pump}} = \frac{\dot{W}_{\text{pump,rev}}}{\dot{W}_{\text{pump}}} \qquad (A.10)$$

where $\dot{W}_{\text{pump,rev}}$ is the reversible pump power, which is equal to $\dot{W}_{\text{pump}} - \dot{I}$. The exergetic efficiencies of a heat exchanger and condenser may be measured by an increase in the exergy of the cold stream divided by the decrease in the exergy of the stream. Applying this definition to a heat exchanger or condenser, we obtain

$$\eta_{\text{ex,heatexc,cond}} = \frac{\left(\dot{E}_{\text{out}} - \dot{E}_{\text{in}}\right)_{\text{cold}}}{\left(\dot{E}_{\text{in}} - \dot{E}_{\text{out}}\right)_{\text{hot}}} \qquad (A.11)$$

where the subscripts cold and hot represent the cold stream and the hot stream, respectively. The difference between the numerator and denominator of (A.11) is the exergy destruction in the heat exchanger or condenser. One may take all the exergy given up by the hot fluid in the condenser as part of the exergy destruction for the power plant. This is the value used in exergy destruction diagram shown later in the appendix.

In general, the thermal efficiency of a geothermal power plant may be expressed as

$$\eta_{\text{th},1} = \frac{\dot{W}_{\text{net,out}}}{\dot{m}_{\text{geo}}\left(h_{\text{geo}} - h_0\right)} \tag{A.12}$$

where the expression in the denominator is the energy input to the power plant, which is expressed as the enthalpy of the geothermal water with respect to the environment state multiplied by the mass flow rate of geothermal water. Using the states, it becomes

$$\eta_{\text{th},1} = \frac{\dot{W}_{\text{net,out}}}{\dot{m}_{\text{geo}}(h_5 - h_0)} \tag{A.13}$$

or according to the energy of geothermal water at the heat exchanger inlet:

$$\eta_{\text{th},2} = \frac{\dot{W}_{\text{net,out}}}{\dot{m}_{\text{geo}}(h_6 - h_0)} \tag{A.14}$$

In (A.13), the energy input to the power plant represents the maximum heat the geothermal water can give and this can only happen when the geothermal water is cooled to the temperature of the environment.

The actual heat input to a geothermal power cycle is less than the term in the denominator of (A.13) because part of the geothermal water is reinjected back to the ground at a temperature much greater than the temperature of the environment. In this approach, the thermal efficiency is determined from

$$\eta_{\text{th},3} = \frac{\dot{W}_{\text{net,out}}}{\dot{Q}_{\text{in}}} \tag{A.15}$$

The thermal efficiency may be expressed based on the heat transfer to the binary Rankine cycle (i.e., the heat transfer in the heat exchanger):

$$\eta_{\text{th,binary}} = \frac{\dot{W}_{\text{net,out}}}{\dot{m}_{\text{geo}}\left(h_6 - h_7\right)} \tag{A.16}$$

or

$$\eta_{\text{th,binary}} = \frac{\dot{W}_{\text{net,out}}}{\dot{m}_{\text{binary}}(h_3 - h_2)} \tag{A.17}$$

Using the exergy of geothermal water as the exergy input to the plant, the exergy efficiency of a geothermal power plant can be expressed as

$$\eta_{ex,1} = \frac{\dot{W}_{net,out}}{\dot{E}_{in}} = \frac{\dot{W}_{net,out}}{\dot{m}_{geo}[h_5 - h_0 - T_0(s_5 - s_0)]} \qquad (A.18)$$

or according to the exergy of geothermal water at the heat exchanger inlet ,

$$\eta_{ex,2} = \frac{\dot{W}_{net,out}}{\dot{E}_6} = \frac{\dot{W}_{net,out}}{\dot{m}_{binary}[h_6 - h_0 - T_0(s_6 - s_0)]} \qquad (A.19)$$

For a binary cycle, the exergy efficiency may be defined based on the exergy decrease of geothermal water or exergy increase of the binary working fluid in the heat exchanger. That is,

$$\eta_{ex,binary,1} = \frac{\dot{W}_{net,out}}{\dot{m}_{geo}[h_6 - h_7 - T_0(s_6 - s_7)]} \qquad (A.20)$$

$$\eta_{ex,binary,2} = \frac{\dot{W}_{net,out}}{\dot{m}_{binary}[h_3 - h_2 - T_0(s_3 - s_2)]} \qquad (A.21)$$

The difference between the denominators of (A.20) and (A.21) is the exergy destruction in the heat exchanger.

The total exergy lost in cycle is determined from

$$\dot{I}_{cycle} = \dot{I}_{pump} + \dot{I}_{heatexchanger} + \dot{I}_{turbine} + \dot{I}_{condenser} + \dot{I}_{reinjection} \qquad (A.22)$$

The total exergy destruction in the plant is the difference between the brine exergy at the heat exchanger inlet and the net power output from the cycle:

$$\dot{I}_{plant} = \dot{E}_{in} - \dot{W}_{net,out} \qquad (A.23)$$

This includes various exergy losses in the plant components as well as the exergy of the brine leaving the heat exchanger. One may argue that the exergy of used brine is a recovered exergy, and it should not be considered to be part of the exergy loss. However, the used brine is reinjected back to the ground without any attempt to make use of it.

The exergetic efficiencies and exergy destruction of major plant components and the entire plant are calculated as explained in this section, and listed in Table A.2. All values are for one representative unit. To pinpoint the sites of exergy destruction and quantify those losses, an exergy diagram is given in Fig. A.4. An energy diagram is given in Fig. A.5 to provide a comparison with the exergy flow diagram.

Table A.2 Some exergetic and energetic performance data provided for one representative unit of the plant

Component	Exergy destruction (kW)	Exergetic efficiency (%)	Heat transfer or power (kW)	Effectiveness or isentropic efficiency (%)
Reinjection well	26,706	–	202,742	–
Heat exchanger	9,552	80.5	160,929	47.1
Air-cooled condenser	14,307	28.2	141,271	88.6
Turbine I	2,411	81.8	10,872	78.2
Turbine II	2,411	81.8	10,872	78.2
Circulation pump	541	74.1	2,087	73.4
Parasitic power			3,262	
Cycle	51,891	41.7 [(Eq. A.21)]	16,396	10.2 [(Eq. A.17)]
		33.5 [(Eq. A.20)]		10.2 [(Eq. A.16)]
		21.7 [(Eq. A.19)]		4.5 [(Eq. A.14)]
		21.1 [(Eq. A.18)]		4.4 [(Eq. A.13)]

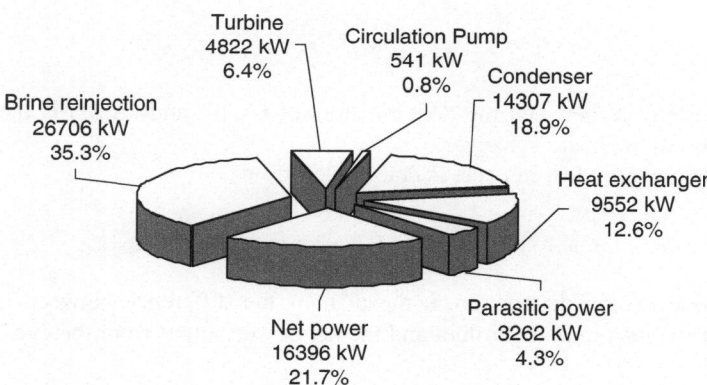

Fig. A.4 Exergy losses diagram. Given as the percentages of brine exergy input (75,586 kW), which is taken as the exergy of brine at state 6 in Fig. A.1

The power output from the turbines is 10,872 kW in Turbine I and 10,872 kW in Turbine II. The pump power requirement for the circulation pumps is calculated to be 2,087 kW. The net power outputs form Rankine cycle then become 19,657 kW. It is further estimated based on plant data about 16.6% of the net power generated in the cycle is consumed by the plant unit parasites, which corresponds to 3,262 kW. Parasitic power includes brine production pumps, condenser fans, and other auxiliaries. Subtracting the parasitic power from the net power generated in the cycle, the net power output becomes 16,396 kW.

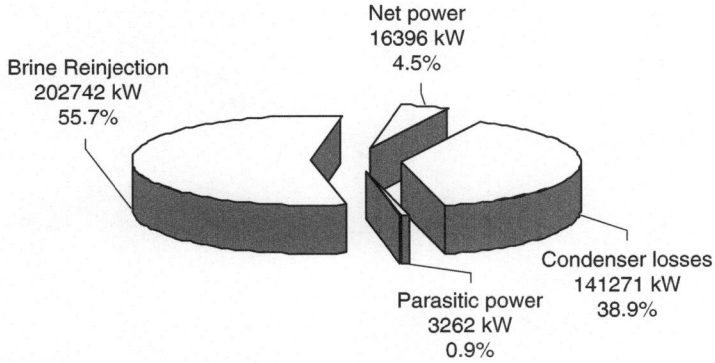

Fig. A.5 Energy losses diagram. Given as the percentages of brine energy input (363,671 kW), which is taken as the energy of brine at state 6 in Fig. A.1

In Table A.1, temperature, pressure, and mass flow rate data for geothermal fluid, working fluid, and air are given according to their state numbers specified in Fig. A.1. Exergy rates are calculated for each state, and listed in Table A.1. States 0, $0'$ and $0''$ are the restricted dead-states for the geothermal fluid, working fluid, and air, respectively. They correspond to an environment temperature of 3°C and an atmospheric pressure of 84 kPa, which were the values measured at the time when the plant data were obtained. For geothermal fluid, the thermodynamic properties of water are used. By doing so, effects of salts and noncondensable gases that might present in the geothermal brine are neglected. This should not cause any significant error in calculations because their fractions are estimated by the plant management to be small. Thermodynamic properties of the working fluid, isobutane, are obtained from EES software with built-in thermodynamic property functions.

As part of the analysis, we investigated the effects of turbine inlet pressure and temperature and the condenser pressure on exergy and energy efficiencies, the net power output, and the brine reinjection temperature. In order to facilitate this parametric study, we used the given geothermal inlet temperature and flow rate values (158°C, 555.9 kg/s) and the calculated isentropic efficiencies for the turbine (0.782) and pump (0.734), and the pinch-point temperature difference (6°C). The brine temperature at the heat exchanger exit (reinjection temperature) and the mass flow rate of isobutane are the unknown parameters in this analysis. The results of this parametric study are given in Figs. A.6 through A.11.

A.1.4 Results and Discussion

An investigation of the exergy pie diagram given in Fig. A.4 shows that 74% of the exergy entering the plant is lost. The remaining 26% is converted to power and 21.7% of this power is used for parasitic load in the plant. The exergetic efficiency

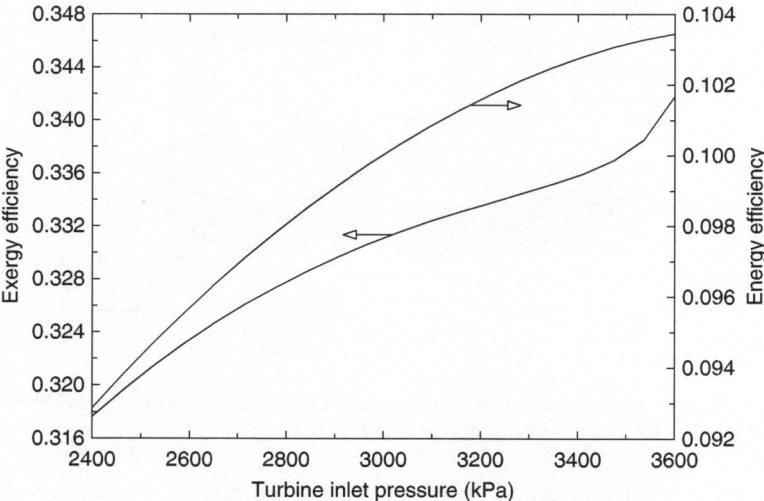

Fig. A.6 Variation of exergy and energy efficiencies versus turbine inlet pressure

of the plant is determined to be 41.7% [(A.21)] and 33.5% [(A.20)] based on the exergy input to the isobutane Rankine cycles (i.e., exergy drops of the brine in the heat exchanger) and 21.1% [(A.18)] and 21.7% [(A.19)] based on the exergy input to the plant (i.e., exergy of the brine at the reservoir and the heat exchanger inlet, respectively; Table A.2).

Using low-temperature resources, geothermal power plants generally have low first law efficiencies. Consequently, the first law efficiency of the plant is calculated to be 4.4% [(A.13)] and 4.5% [(A.14)] based on the energy input to the plant and 10.2% [(A.17)] based on the energy input to the isobutane Rankine cycles. This means that more than 90% of the energy of the brine in the reservoir is discarded as waste heat. There is a strong argument here for the use of geothermal resources for direct applications such as district heating instead of power generation when economically feasible. A cogeneration scheme involving power generation and district heating may also be considered when used brine is reinjected back to the ground at a relatively high temperature. The energy losses diagram in Fig. A.5 shows that 55.7% of the energy of the brine is reinjected, 38.9% of it is rejected in the condenser, and the remaining is converted to power. Yet it provides no specific information on where the power potentials are lost. This shows the value of an exergy analysis.

The causes of exergy destruction in the plant include heat exchanger losses, turbine-pump losses, the exergy of the brine reinjected, and the exergy of isobutane lost in the condenser. They represent 18.9%, 7.2%, 35.3%, and 12.6% of the brine exergy input, respectively (Fig. A.4). The exergetic efficiencies and effectiveness of heat exchanger are 80.5% and 47.1%, respectively. This exergetic efficiency can be considered to be high, and indicate a satisfactory performance of the heat

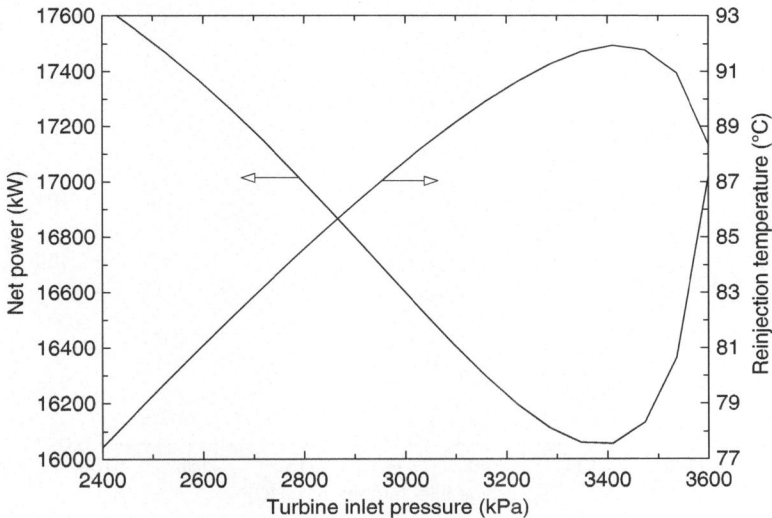

Fig. A.7 Variation of net power and brine reinjection temperature versus turbine inlet pressure

exchange system. In binary geothermal power plants heat exchangers are important components and their individual performances affect the overall performance of the plant considerably.

The exergetic efficiency of the turbine is 81.8%, which is reasonable. The exergetic efficiencies of the condensers are in the range of 28.2%, making them the least efficient components in the plant. This is primarily due to the high average temperature difference between the isobutane and the cooling air. The brine is reinjected back to the ground at about 90°C.

For binary geothermal power plants using air as the cooling medium, the condenser temperature varies as the ambient air temperature fluctuates throughout the year and even throughout the day. As a result, the power output decreases by up to 50% from winter to summer. Consequently, the exergy destruction rates and percentages at various sites change, this effect being most noticeable in the condenser.

As part of the analysis, we investigate the effect of some operating parameters on exergy and energy efficiencies, the net power output, and the brine reinjection temperature. The energy and exergy efficiencies are those given in (A.16) and (A.20), respectively. The effect of turbine inlet pressure on the exergy and energy efficiencies is given in Fig. A.6. Both the exergy and energy efficiencies increase with the turbine inlet pressure. The critical pressure of isobutane is 3,640 kPa. As the pressure approaches the critical pressure the increase in exergy efficiency becomes more dramatic. On the other hand, energy efficiency shows a different trend as the pressure approaches the critical value. Figure A.7 indicates a pressure of about 3,430 kPa at which the net power from the plant is minimized. This is also about the same pressure at which the brine reinjection temperature is a maximum.

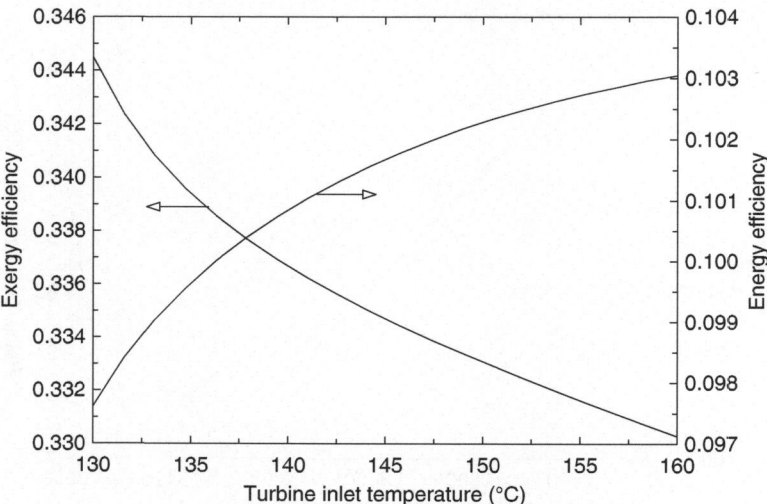

Fig. A.8 Variation of exergy and energy efficiencies versus turbine inlet temperature

This means that less heat is picked up from the geothermal brine by the isobutane. For geothermal power plants where the brine leaving the heat exchanger is used for no useful purpose and directly reinjected to the ground, maximizing the power output (not the energy or exergy efficiency) is a priority. This is the case for this particular power plant. Note that the brine temperature is high enough for use in district heating systems. This may be explored if there is a residential, commercial, or industrial district within close proximity to the power plant.

The exergy efficiency decreases and the energy efficiency increases with increasing turbine inlet temperature as shown in Fig. A.8. The reason for the decreasing exergy efficiency trend is this: as the temperature increases the power potential increases but the power output decreases (Fig. A.9). The reason for the decreasing trend in the power output is the decrease in mass flow rate of isobutane. It decreases from 457 kg/s at 130°C to 266 kg/s at 155°C. The reason for the increasing trend in energy efficiency is because both the power output and the heat input decrease with increasing turbine inlet temperature whereas heat input decreases at a greater rate than power output. The reason for decreasing heat input is due to the decreasing mass flow of isobutane.

Figures A.10 and A.11 show that exergy and energy efficiencies and the net power decrease as the condenser pressure increases. The reinjection temperature remains almost constant with varying condenser pressure. Note that the mass flow rate of isobutane and the rate of heat input to the Rankine cycle remains essentially constant when condenser pressure is changed because changing condenser pressure only affects part of the heat exchange process described in (A.2) whose effect is very small.

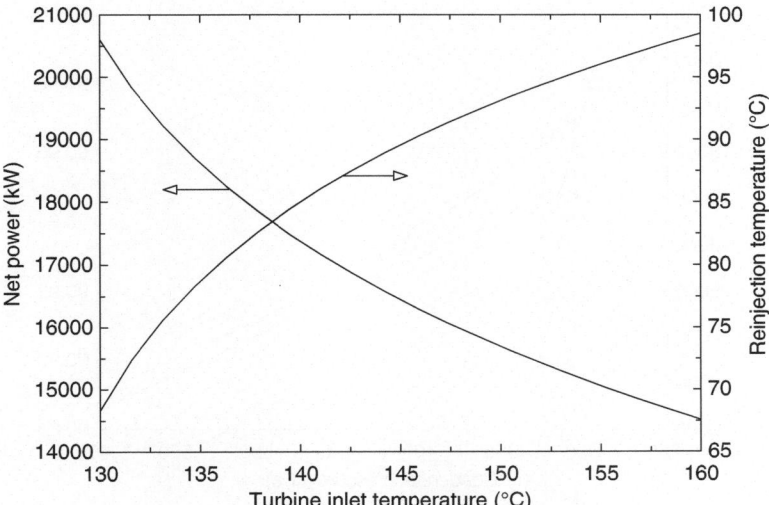

Fig. A.9 Variation of net power and brine reinjection temperature versus turbine inlet temperature

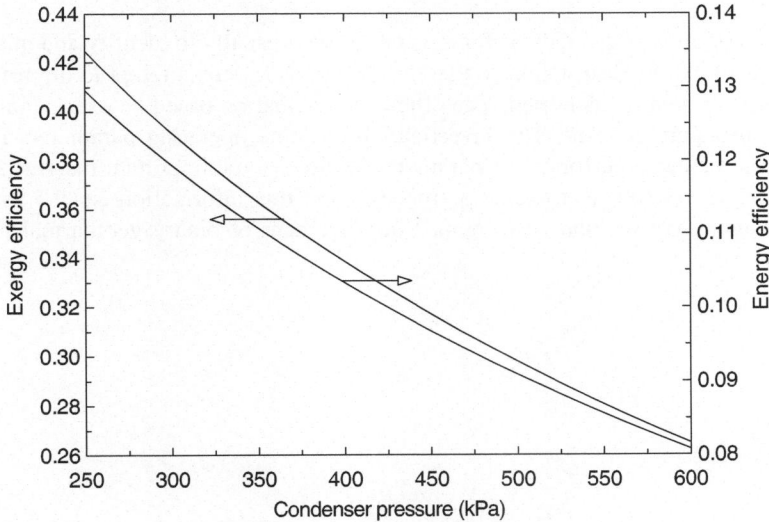

Fig. A.10 Variation of exergy and energy efficiencies versus condenser pressure

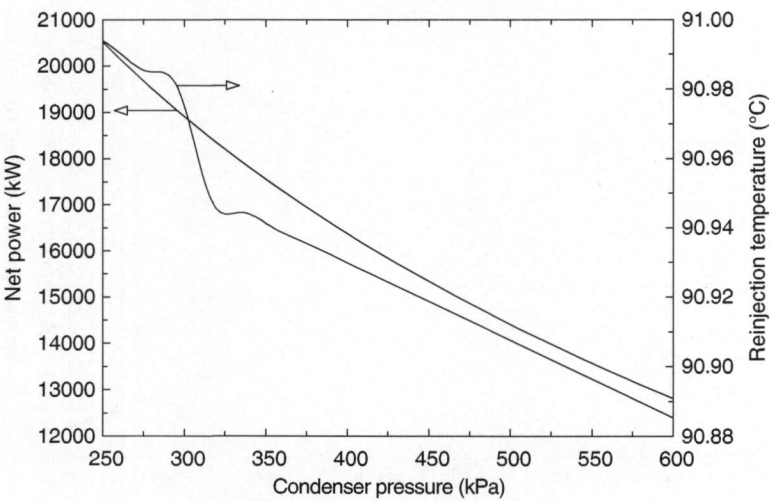

Fig. A.11 Variation of net power and brine reinjection temperature versus condenser pressure

Conclusions

The aim of the exergy analysis for a power plant is usually to identify and quantify the sites of exergy destruction so that the directions for any attempt to improve the performance can be identified. The efficiency evaluation based on exergy analysis serves this purpose well. The investigation of some operating parameters in the cycle on the cycle performance parameters yielded some important insights to the plant operation and heat exchange process, and this information can be used in the design, analysis, and performance improvement of binary geothermal power plants.

Appendix B

This case study is based on the study by

M. Kanoğlu, Exergy analysis of multistage cascade refrigeration cycle used for natural gas liquefaction, in the *International Journal of Energy Research,* 26: 763–774, 2002.

B.1 A Case Study on Efficiency Evaluation of a Multistage Cascade Refrigeration Cycle for Natural Gas Liquefaction

B.1.1 Introduction

Natural gas is a mixture of components, consisting mainly of methane (60–98%) with small amounts of other hydrocarbon fuel components. It also contains various amounts of nitrogen, carbon dioxide, helium, and traces of other gases. It is stored as compressed natural gas (CNG) at pressures of 16–25 MPa and around room temperature, or as a liquefied natural gas (LNG) at pressures of 70–500 kPa and around −150°C or lower. When transportation of natural gas in pipelines is not feasible for economic and other reasons, it is first liquefied using unconventional refrigeration cycles and then it is usually transported by marine ships in specially made insulated tanks. It is regasified in receiving stations before given off the pipeline for end use. In fact, different refrigeration cycles with different refrigerants can be used for natural gas liquefaction.

The first (and still commonly) used cycle for natural gas liquefaction was the multistage cascade refrigeration cycle that uses three different refrigerants, namely propane, ethane (or ethylene), and methane in their individual refrigeration cycles. A great amount of work is consumed to obtain LNG at about −150°C which enters the cycle at about atmospheric temperature in the gas phase. Minimizing the work

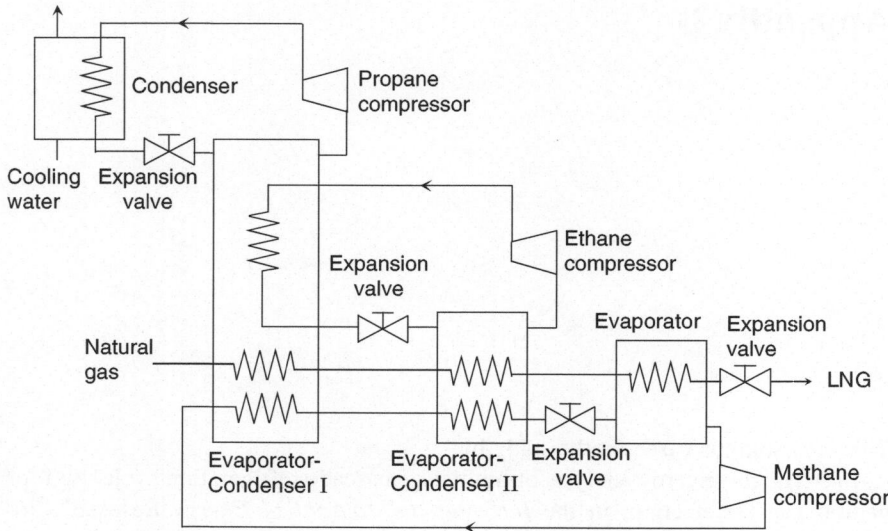

Fig. B.1 Schematic of the cascade refrigeration cycle (showing only one stage for each refrigerant cycle for simplicity)

consumed in the cycle is the most effective measure to reduce the cost of LNG. In this regard, exergy appears to be a potential tool for the design, optimization, and performance evaluation of such systems. Note that identifying the main sites of exergy destruction shows the direction for potential improvements. An important object of exergy analysis for systems that consume work such as liquefaction of gases and distillation of water is finding the minimum work required for a certain desired result.

B.1.2   Description of the Cycle

Figure B.1 shows a schematic of the cascade refrigeration cycle and its components. The cycle consists of three subcycles and each one uses a different refrigerant. In the first cycle, propane leaves the compressor at a high temperature and pressure and enters the condenser where the cooling water or air is used as the coolant. The condensed propane then enters the expansion valve where its pressure is decreased to the evaporator pressure. As the propane evaporates, the heat of evaporation comes from the condensing ethane, cooling methane, and cooling natural gas. Propane leaves the evaporator and enters the compressor, thus completing the cycle. The condensed ethane expands in the expansion valve and evaporates as methane condenses and natural gas is further cooled and liquefied. Finally, methane expands and then evaporates as natural gas is liquefied and subcooled. As methane enters the compressor to complete the cycle, the LNG pressure is dropped in an expansion valve to the storage pressure. The three refrigerant cycles have multistage compression and expansion, usually with three stages, and consequently three evaporation temperature levels for each refrigerant. The mass flows in each stage are usually different. Natural gas from the pipeline goes through a process during which the acid gases are removed and its pressure is increased to an average value of 40 bar before entering the cycle.

B.1.3   Exergy Analysis

The flow exergy of any fluid in a control volume can be written as follows (with the negligible changes in kinetic and potential energies).

$$\dot{X} = \dot{m}e = \dot{m}[(h - h_0) - T_0(s - s_0)] \tag{B.1}$$

where T_0 is the dead-state temperature, h and s are the enthalpy and entropy of the fluid at the specified state, and h_0 and s_0 are the corresponding properties at the dead- (reference) state.

The specific exergy change between two states (e.g., inlet and outlet) is

$$\Delta e = e_1 - e_2 = (h_1 - h_2) - T_0(s_1 - s_2) \tag{B.2}$$

As mentioned earlier, some part of the specific exergy change is lost during the process due to entropy generation; referring to $T_0\Delta s$ for the above equation, this is $i = T_0\Delta s = T_0 s_{gen}$ known as specific irreversibility. Here s_{gen} is the entropy generation. Two main causes for entropy generation are friction and heat transfer across a finite temperature difference. Heat transfer is always accompanied by exergy transfer, which is given by

$$e_q = \int \delta q \left(1 - \frac{T_0}{T}\right) \tag{B.3}$$

where δq is the differential heat transfer and T is the source temperature where heat transfer takes place. Heat transfer is assumed to occur with the surroundings at T_0. If this heat transfer shows an undesired heat loss, (B.3) also expresses the exergy lost by heat.

The following are the exergy destruction and exergetic efficiency relations for various cycle components as shown in Fig. B.1.

B.1.3.1 Evaporators and Condensers

The evaporators and condensers in the system are treated as heat exchangers. There are a total of four evaporator–condenser systems in the cycle. The first system, named evaporator–condenser-I, is the propane cycle evaporator and the ethane and methane cycle condenser. Similarly, the system named evaporator–condenser-II is the ethane cycle evaporator and the methane cycle condenser. The third system is the methane cycle evaporator and the fourth system is the propane cycle condenser where the cooling water is used as coolant. An exergy balance written on the evaporator–condenser I should express the exergy loss in the system as the difference of exergies of incoming and outgoing streams. That is,

$$
\dot{I} = \dot{E}_{\text{in}} - \dot{E}_{\text{out}} = \left[\sum (\dot{m}_p e_p) + \sum (\dot{m}_e e_e) + \sum (\dot{m}_m e_m) + (\dot{m}_d e_d) \right]_{\text{in}}
$$
$$
- \left[\sum (\dot{m}_p e_p) + \sum (\dot{m}_e e_e) + \sum (\dot{m}_m e_m) + (\dot{m}_n e_n) \right]_{\text{out}}
\tag{B.4}
$$

where the subscripts in, out, p, e, m, and n stand for inlet, outlet, propane, ethane, methane, and natural gas, respectively. The summation signs are due to the fact that there are three stages in each refrigerant cycle with different pressures, evaporation temperatures, and mass flow rates.

The exergetic efficiency of a heat exchanger can be defined as the ratio of total outgoing stream exergies to total incoming stream exergies as follows.

$$
\varepsilon = \frac{\sum (\dot{m}_p e_p)_{\text{out}} + \sum (\dot{m}_e e_e)_{\text{out}} + \sum (\dot{m}_m e_m)_{\text{out}} + (\dot{m}_n e_n)_{\text{out}}}{\sum (\dot{m}_p e_p)_{\text{in}} + \sum (\dot{m}_e e_e)_{\text{in}} + \sum (\dot{m}_m e_m)_{\text{in}} + (\dot{m}_n e_n)_{\text{in}}}
\tag{B.5}
$$

The second definition for the exergy efficiency of heat exchangers can be the ratio of the increase in the exergy of the cold fluid to the decrease in the exergy of the hot. In the system, the only fluid with an exergy increase is propane whereas the exergies of ethane, methane, and natural gas decrease. Therefore, the equation becomes:

$$
\varepsilon = \frac{\sum (\dot{m}_p e_p)_{\text{out}} - \sum (\dot{m}_p e_p)_{\text{in}}}{\sum (\dot{m}_e e_e)_{\text{in}} - \sum (\dot{m}_e e_e)_{\text{out}} + \sum (\dot{m}_m e_m)_{\text{in}} - \sum (\dot{m}_m e_m)_{\text{out}} + (\dot{m}_n e_n)_{\text{in}} - (\dot{m}_n e_n)_{\text{out}}}
$$
$$
\tag{B.6}
$$

The above two methods used to determine the exergetic efficiency of a heat exchanger are sometimes called the scientific approach and the engineering approach, respectively. The efficiencies calculated using these two approaches are usually very close to each other. Here in this example, the engineering approach is used in the following relations. The relations for exergy destruction and exergetic efficiency for evaporator–condenser II are determined as

$$I = \dot{E}_{in} - \dot{E}_{out} = \left[\sum (\dot{m}_e e_e) + \sum (\dot{m}_m e_m) + (\dot{m}_n e_n)\right]_{in}$$
$$- \left[\sum (\dot{m}_e e_e) + \sum (\dot{m}_m e_m) + (\dot{m}_n e_n)\right]_{out} \quad \text{(B.7)}$$

$$\varepsilon = \frac{\sum (\dot{m}_e e_e)_{out} - \sum (\dot{m}_e e_e)_{in}}{\sum (\dot{m}_m e_m)_{in} - \sum (\dot{m}_m e_m)_{out} + (\dot{m}_n e_n)_{in} - (\dot{m}_n x_n)_{out}} \quad \text{(B.8)}$$

From the exergy balance on the evaporator of the methane cycle the following exergy destruction and exergetic efficiency expressions can be written.

$$I = \dot{E}_{in} - \dot{E}_{out} = \left[\sum (\dot{m}_m e_m) + (\dot{m}_n e_n)\right]_{in} - \left[\sum (\dot{m}_m e_m) + (\dot{m}_n e_n)\right]_{out} \quad \text{(B.9)}$$

$$\varepsilon = \frac{\sum (\dot{m}_m e_m)_{out} - \sum (\dot{m}_m e_m)_{in}}{(\dot{m}_n e_n)_{in} - (\dot{m}_n e_n)_{out}} \quad \text{(B.10)}$$

Finally, for the condenser of the propane cycle the following can be obtained,

$$I = \dot{E}_{in} - \dot{E}_{out} = \left[\sum (\dot{m}_p e_p) + (\dot{m}_w e_w)\right]_{in} - \left[\sum (\dot{m}_p e_p) + (\dot{m}_w e_w)\right]_{out} \quad \text{(B.11)}$$

$$\varepsilon = \frac{(\dot{m}_w e_w)_{out} - (\dot{m}_w e_w)_{in}}{\sum (\dot{m}_p e_p)_{in} - \sum (\dot{m}_p e_p)_{out}} \quad \text{(B.12)}$$

where the subscript w stands for water.

B.1.3.2 Compressors

There is one multistage compressor in the cycle for each refrigerant. The total work consumed in the cycle is the sum of work input to the compressors. There is no exergy destruction in a compressor if irreversibilities can be totally eliminated. This results in a minimum work input for the compressor. In reality, there are irreversibilities due to friction, heat loss, and other dissipative effects. The exergy destruction in propane, ethane, and methane compressors can be expressed, respectively, as

$$I_p = \dot{E}_{in} - \dot{E}_{out} = \sum (\dot{m}_p e_p)_{in} + \dot{W}_{p,in} - \sum (\dot{m}_p e_p)_{out} \quad \text{(B.13)}$$

$$I_e = \dot{E}_{in} - \dot{E}_{out} = \sum (\dot{m}_e e_e)_{in} + \dot{W}_{e,in} - \sum (\dot{m}_e e_e)_{out} \quad \text{(B.14)}$$

$$\dot{I}_m = \dot{E}_{in} - \dot{E}_{out} = \sum (\dot{m}_m e_m)_{in} + \dot{W}_{m,in} - \sum (\dot{m}_m e_m)_{out} \qquad (B.15)$$

where $\dot{W}_{p,in}, \dot{W}_{e,in}$, and $\dot{W}_{m,in}$ are the actual power inputs to the propane, ethane, and methane compressors, respectively. They are part of the exergy inputs to the compressors. The exergetic efficiency of the compressor can be defined as the ratio of the minimum work input to the actual work input. The minimum work is simply the exergy difference between the actual inlet and exit states. Applying this definition to propane, ethane, and methane compressors, respectively, the exergy efficiency equations become

$$\varepsilon_p = \frac{\sum (\dot{m}_p e_p)_{out} - \sum (\dot{m}_p e_p)_{in}}{\dot{W}_{p,in}} \qquad (B.16)$$

$$\varepsilon_e = \frac{\sum (\dot{m}_e e_e)_{out} - \sum (\dot{m}_e e_e)_{in}}{\dot{W}_{e,in}} \qquad (B.17)$$

$$\varepsilon_m = \frac{\sum (\dot{m}_m e_m)_{out} - \sum (\dot{m}_m e_m)_{in}}{\dot{W}_{m,in}} \qquad (B.18)$$

B.1.3.3 Expansion Valves

Beside the expansion valves in the refrigeration cycles, one is used to drop the pressure of LNG to the storage pressure. Expansion valves are considered essentially isenthalpic devices with no work interaction and negligible heat transfer with the surroundings. From an exergy balance, the exergy destruction equations for propane, ethane, methane, and LNG expansion valves can be written as

$$\dot{I}_p = \dot{E}_{in} - \dot{E}_{out} = \sum (\dot{m}_p e_p)_{in} - \sum (\dot{m}_p e_p)_{out} \qquad (B.19)$$

$$\dot{I}_e = \dot{E}_{in} - \dot{E}_{out} = \sum (\dot{m}_e e_e)_{in} - \sum (\dot{m}_e e_e)_{out} \qquad (B.20)$$

$$\dot{I}_m = \dot{E}_{in} - \dot{E}_{out} = \sum (\dot{m}_m e_m)_{in} - \sum (\dot{m}_m e_m)_{out} \qquad (B.21)$$

$$\dot{I}_n = \dot{E}_{in} - \dot{E}_{out} = \sum (\dot{m}_n e_n)_{in} - \sum (\dot{m}_n e_n)_{out} \qquad (B.22)$$

The exergetic efficiency of expansion valves can be defined as the ratio of the total exergy output to the total exergy input. Therefore, the exergy efficiencies for all expansion valves become

$$\varepsilon_p = \frac{\sum (\dot{m}_p e_p)_{out}}{\sum (\dot{m}_p e_p)_{in}} \qquad (B.23)$$

$$\varepsilon_e = \frac{\sum (\dot{m}_e e_e)_{\text{out}}}{\sum (\dot{m}_e e_e)_{\text{in}}} \tag{B.24}$$

$$\varepsilon_m = \frac{\sum (\dot{m}_m e_m)_{\text{out}}}{\sum (\dot{m}_m e_m)_{\text{in}}} \tag{B.25}$$

$$\varepsilon_n = \frac{\sum (\dot{m}_n e_n)_{\text{out}}}{\sum (\dot{m}_n e_n)_{\text{in}}} \tag{B.26}$$

B.1.3.4 Cycle

The total exergy destruction in the cycle is simply the sum of exergy destruction in condensers, evaporators, compressors, and expansion valves. This total can be obtained by adding the exergy destruction terms in the above equations. Then, the overall exergy efficiency of the cycle can be defined as

$$\varepsilon = \frac{\dot{E}_{\text{out}} - \dot{E}_{\text{in}}}{\dot{W}_{\text{actual}}} = \frac{\dot{W}_{\text{actual}} - \dot{I}_{\text{total}}}{\dot{W}_{\text{actual}}} \tag{B.27}$$

where given in the numerator is the exergy difference or the actual work input to the cycle $\dot{W}_{\text{actual}}$ minus the total exergy destruction $\dot{I}$. The actual work input to the cycle is the sum of the work inputs to the propane, ethane, and methane compressors, as follows.

$$\dot{W}_{\text{actual}} = \dot{W}_{p,\text{in}} + \dot{W}_{e,\text{in}} + \dot{W}_{m,\text{in}} \tag{B.28}$$

In this regard, the exergetic efficiency of the cycle can also be expressed as

$$\varepsilon = \frac{\dot{W}_{\text{min}}}{\dot{W}_{\text{actual}}} \tag{B.29}$$

where $\dot{W}_{\text{min}}$ is the minimum work input to the cycle. Here, a process is proposed to determine the minimum work input to the cycle, or in other words, the minimum work for liquefaction process.

The exergetic efficiency of the natural gas liquefaction process can be defined as the ratio of the minimum work required to produce a certain amount of LNG to the actual work input. An exergy analysis needs to be performed on the cycle to determine the minimum work input. The liquefaction process is essentially the removal of heat from the natural gas. Therefore, the minimum work can be determined by utilizing a reversible or Carnot refrigerator. The minimum work

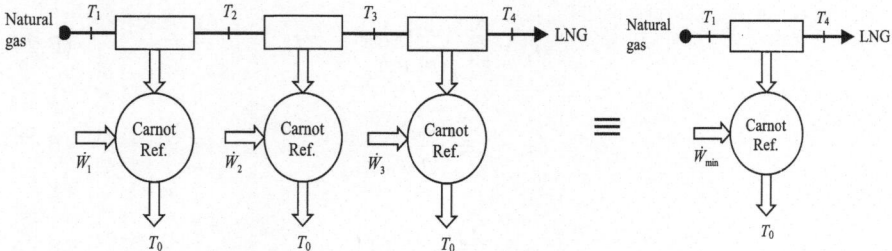

Fig. B.2 Determination of minimum work for the cycle

input for the liquefaction process is simply the work input required for the operation of a Carnot refrigerator for a given heat removal. It can be expressed as

$$w_{\min} = \int \delta q \left(1 - \frac{T_0}{T} \right) \tag{B.30}$$

where δq is the differential heat transfer and T is the instantaneous temperature at the boundary where the heat transfer takes place. Note that T is smaller than T_0 for the liquefaction process and to get a positive work input we have to take the sign of heat transfer to be negative inasmuch as it is a heat output. The evaluation of (B.30) requires knowledge of the functional relationship between the heat transfer δq and the boundary temperature T, which is usually not available.

As seen in Fig. B.1, natural gas flows through three evaporator–condenser systems in the multistage refrigeration cycle before it is fully liquefied. Thermodynamically, this three-stage heat removal from natural gas can be accomplished using three Carnot refrigerators as seen in Fig. B.2. The first Carnot refrigerator receives heat from the natural gas and supplies it to the heat sink at T_0 as the natural gas is cooled from T_1 to T_2. Similarly, the second Carnot refrigerator receives heat from the natural gas and supplies it to the heat sink at T_0 as the natural gas is cooled from T_2 to T_3. Finally, the third Carnot refrigerator receives heat from the natural gas and supplies it to the heat sink at T_0 as the natural gas is further cooled from T_3 to T_4, where it exists as LNG. The amount of power that needs to be supplied to each Carnot refrigerator can be determined from

$$\dot{W}_{\min} = \dot{W}_1 + \dot{W}_2 + \dot{W}_3 = \dot{m}_n(e_1 - e_4) = \dot{m}_n[h_1 - h_4 - T_0(s_1 - s_4)] \tag{B.31}$$

where $\dot{W}_1$, $\dot{W}_2$, and $\dot{W}_3$ are the power inputs to the first, second, and third Carnot refrigerators, respectively:

$$\dot{W}_1 = \dot{m}_n(e_1 - e_2) = \dot{m}_n[h_1 - h_2 - T_0(s_1 - s_2)] \tag{B.32}$$

$$\dot{W}_2 = \dot{m}_n(e_2 - e_3) = \dot{m}_n[h_2 - h_3 - T_0(s_2 - s_3)] \qquad (B.33)$$

$$\dot{W}_3 = \dot{m}_n(e_3 - e_4) = \dot{m}_n[h_3 - h_4 - T_0(s_3 - s_4)] \qquad (B.34)$$

This is the expression for the minimum power input for the liquefaction process. This minimum power can be obtained by using a single Carnot refrigerator that receives heat from the natural gas and supplies it to the heat sink at T_0 as the natural gas is cooled from T_1 to T_4. That is, this Carnot refrigerator is equivalent to the combination of three Carnot refrigerators as shown in Fig. B.2. The minimum work required for the liquefaction process depends only on the properties of the incoming and outgoing natural gas and the ambient temperature T_0.

B.1.3.5   Illustrative Example

We use numerical values to study the multistage cascade refrigeration cycle used for natural gas liquefaction. A numerical value of the minimum work can be calculated using typical values of incoming and outgoing natural gas properties. The pressure of natural gas is around 40 bar when entering the cycle. The temperature of natural gas at the cycle inlet can be taken to be the same as the ambient temperature $T_1 = T_0 = 25°C$. Natural gas leaves the cycle liquefied at about 4 bar pressure and $-150°C$ temperature. The natural gas in the cycle usually consists of more than 95% methane, therefore thermodynamic properties for methane can be used for natural gas. Using these inlet and exit states, the minimum work input to produce a unit mass of LNG can be determined from (B.31) to be 456.8 kJ/kg. The heat removed from the natural gas during the liquefaction process is determined from

$$\dot{Q} = \dot{m}_n(h_1 - h_4) \qquad (B.35)$$

For the inlet and exit states of natural gas described above, the heat removed from the natural gas can be determined from (B.35) to be 823.0 kJ/kg. That is, for the removal of 823.0 kJ/kg heat from the natural gas, a minimum of 456.8 kJ/kg work is required. Because the ratio of heat removed to the work input is defined as the COP of a refrigerator, this corresponds to a COP of 1.8. That is, the COP of the Carnot refrigerator used for natural gas liquefaction is only 1.8. This is expected due to the high difference between the temperature T and T_0 in (B.30). An average value of T can be obtained from the definition of the COP for a Carnot refrigerator, which is expressed as

$$COP_{R,rev} = \frac{1}{T_0/T - 1} \qquad (B.36)$$

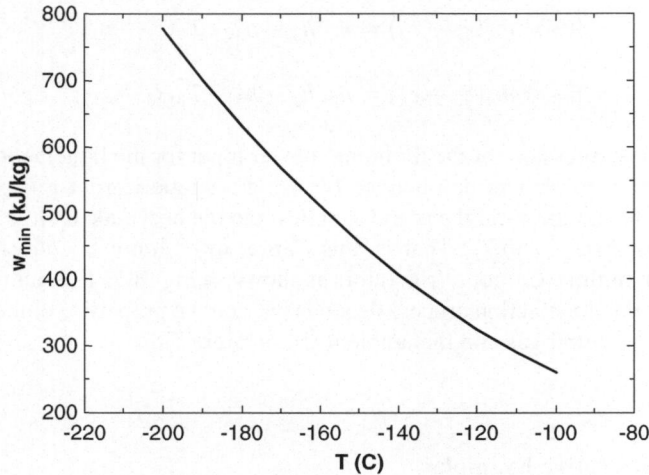

Fig. B.3 Minimum work ($w_{\min}$) versus natural gas liquefaction temperature

Using this equation, for COP $= 1.8$ and $T_0 = 25°C$ we determine $T = -81.3°C$. This is the temperature a heat reservoir would have if a Carnot refrigerator with a COP of 1.8 operated between this reservoir and another reservoir at $25°C$. Note that the same result could be obtained by writing (B.30) in the form

$$w_{\min} = q\left(1 - \frac{T_0}{T}\right) \tag{B.37}$$

where $q = 823.0$ kJ/kg, $w_{\min} = 456.8$ kJ/kg, and $T_0 = 25°C$.

As part of the analysis we now investigate how the minimum work changes with the natural gas liquefaction temperature. We take the inlet pressure of natural gas to be 40 bar, the inlet temperature to be $T_1 = T_0 = 25°C$, and the exit state to be the saturated liquid at the specified temperature. The properties of methane are obtained from thermodynamic tables. Using the minimum work relation in (B.37), the plot shown in Fig. B.3 is obtained. Using (B.36), the variation of COP of the Carnot refrigerator with the natural gas liquefaction temperature is also obtained and shown in Fig. B.4.

As shown in the figure, the minimum work required to liquefy a unit mass of natural gas increases almost linearly with the decreasing liquefaction temperature. Obtaining LNG at $-200°C$ requires exactly three times the minimum work required to obtain LNG at $-100°C$. Similarly, obtaining LNG at $-150°C$ requires exactly 1.76 times the minimum work required to obtain LNG at $-100°C$. The COP of the

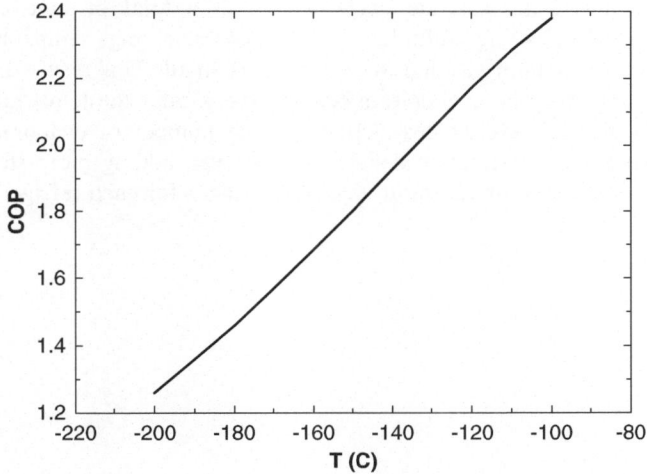

Fig. B.4 COP versus natural gas liquefaction temperature

Carnot refrigerator decreases almost linearly with the decreasing liquefaction temperature as shown in Fig. B.4. The COP decreases almost by half when the liquefaction temperature decreases from −100°C to −200°C. These figures show that the maximum possible liquefaction temperature should be used to minimize the work input. In another words, the LNG should not be liquefied to lower temperatures than needed.

For a typical natural gas inlet and exit states specified in the previous section, the minimum work is determined to be 456.8 kJ/kg of LNG. A typical actual value of work input for a cascade cycle used for natural gas liquefaction may be 1,188 kJ/kg of LNG. Then the exergetic efficiency of a typical cascade cycle can be determined to be 38.5%. The actual work input required depends mainly on the feed and ambient conditions, and on the compressor efficiency.

It has been possible to replace the JT valve of the cycle with a cryogenic hydraulic turbine. The same pressure drop as in a JT valve is achieved with the turbine while producing power. Using the same typical values as taken before, the cryogenic turbine inlet state is 40 bar and −150°C. Assuming isentropic expansion to a pressure of 4 bar, the work output is calculated to be 8.88 kJ/kg of LNG. This corresponds to a decrease of 2% in the minimum work input.

Note that the main site of exergy destruction in the cycle is the compressors. Any improvement in the exergetic efficiency of the compressors will automatically yield lower work input for the liquefaction process. Having three-stage evaporation for each refrigerant in the cascade cycle results in a total of nine evaporation temperatures. Also, having multiple stages makes the average temperature difference between the natural gas and the refrigerants small. This results in a smaller exergy destruction in the evaporators because the greater the temperature difference, the greater the exergy destruction. As the number of evaporation stages increases the exergy destruction decreases. However, adding more stages means additional equipment cost and more than three stages for each refrigerant are not justified.

Nomenclature

c_p	Specific heat at constant pressure (kJ/kg · K)
c_v	Specific heat at constant volume (kJ/kg · K)
COP	Coefficient of performance
E	Energy (kJ)
EER	Energy efficiency ratio
g	Gravitational acceleration (m/s^2)
h	Enthalpy (kJ/kg)
h_{fg}	Enthalpy of vaporization (kJ/kg)
HV	Heating value (kJ/kg)
HHV	Higher heating value (kJ/kg)
I	Current (amp)
k	Specific heat ratio
KE	Kinetic energy (kJ)
LHV	Lower heating value (kJ/kg)
m	Mass (kg)
$\dot{m}$	Mass flow rate (kg/s)
n	Polytropic constant
P	Pressure (kPa)
PE	Potential energy (kJ)
PER	Primary energy ratio
q	Specific heat transfer (kJ/kg)
r	Compression ratio
r_c	Cutoff ratio
r_p	Pressure ratio
R	Gas constant (kJ/kg · K)
Q	amount of heat transfer (kJ)
$\dot{Q}$	Rate of heat transfer (kW)
s	Specific entropy (kJ/kg · K)
S	Total entropy (kJ/K)
S_{gen}	Entropy generation (kJ/K)

SEER	Seasonal energy efficiency ratio
t	Time (s)
T	Temperature (K or °C)
u	Specific internal energy (kJ/kg)
U	Total internal energy (kJ)
v	Specific volume (m³/kg)
V	Velocity (m/s)
V	Voltage (V)
V	Volume (m³)
$\dot{V}$	Volume flow rate (m³/s)
W	Amount of work (kJ)
$\dot{W}$	Rate of work or power (kW)
x	Specific exergy (kJ/kg)
X	Amount of exergy (kJ)
$X_{\text{destroyed}}$	Exergy destruction (kJ)
$\dot{X}$	Rate of exergy (kW)
z	Elevation (m)

Greek Letters

ε	Exergy efficiency
η	Energy efficiency
η_{th}	Thermal efficiency
ϕ	Nonflow exergy (kJ/kg)
ψ	Flow exergy (kJ/kg)

Subscripts

0	Dead (environmental) state
amb	Ambient
Comp	Compressor
Cond	Condenser
CV	Control volume
e	Electricity
elect	Electricity
Evap	Evaporator
Exp Valve	Expansion valve
H	High temperature
HP	Heat pump
in	Inlet
isen	Isentropic
L	Low temperature
mech	Mechanical
out	Outlet
P	Pump

R	Refrigerator
regen	Regenerator
rev	Reversible
s	Source, isentropic
surr	Surroundings
th	Thermal
w	Water

References

1. Y.A. Cengel, M.A. Boles, *Thermodynamics: An Engineering Approach*, 7th edn. (McGraw-Hill, New York, 2011)
2. I. Dincer, M.A. Rosen, Thermodynamic aspects of renewables and sustainable development. Renew. Sust. Energ. Rev. **9**, 169–189 (2005)
3. M.A. Rosen, I. Dincer, M. Kanoglu, Role of exergy in increasing efficiency and sustainability and reducing environmental impact. Energ. Policy **36**, 128–137 (2008)
4. L. Connelly, C.P. Koshland, Two aspects of consumption: using an exergy-based measure of degradation to advance the theory and implementation of industrial ecology. Resour. Conserv. Recycl. **19**, 199–217 (1997)
5. T.J. Kotas, *The Exergy Method in Thermal Plant Analysis*, 2nd edn. (Krieger, Malabar, 1995)
6. M. Kanoglu, I. Dincer, M.A. Rosen, Understanding energy and exergy efficiencies for improved energy management in power plants. Energ. Policy **35**, 3967–3978 (2007)
7. R.L. Cornelissen, Thermodynamics and sustainable development: the use of exergy analysis and the reduction of irreversibility. Ph.D. thesis, University of Twente, 1997
8. A. Bejan, *Advanced Engineering Thermodynamics*, 3rd edn. (Wiley, New York, 2006)
9. I. Dincer, M.A. Rosen, *Exergy, Energy, and Sustainable Development* (Elsevier, Boston, 2007)
10. J. Szargut, D.R. Morris, F.R. Steward, *Exergy Analysis of Thermal, Chemical, and Metallurgical Processes* (Hemisphere Publishing, New York, 1988)
11. K. Wark, *Advanced Thermodynamics for Engineers* (McGraw-Hill, New York, 1995)
12. I. Dincer, The role of exergy in energy policy making. Energ. Policy **30**, 137–149 (2002)
13. R.L. Cornelissen, G.G. Hirs, T.J. Kotas, Different definitions of exergetic efficiencies, in *Proceedings of JETC IV* (Nancy, France, 1995), pp. 335–344
14. H. Struchtrup, M.A. Rosen, How much work is lost in an irreversible turbine? Exergy Int. J. **2**, 152–158 (2002)
15. Y.A. Cengel, H. Kimmel, Power recovery through thermodynamic expansion of liquid methane, in *Proceedings of the American Power Conference*, vol. 59-I, 59th Annual Meeting (Chicago, 1997), pp. 271–276
16. M. Kanoglu, Cryogenic turbine efficiencies. Exergy Int. J. **1**(3), 202–208 (2001)
17. E. Logan, *Handbook of Turbomachinery* (Marcel Dekker, New York, 1995)
18. N. Baines, *New Developments in Radial Turbine Technology* (Imperial College, United Kingdom Concepts ETI, Norwich, 1994)
19. F.M. White, *Fluid Mechanics*, 3rd edn. (McGraw-Hill, New York, 1994)
20. S.A. Klein, Engineering equation solver (EES), F-Chart Software (2006), www.fChart.com
21. M. Kanoglu, Thermodynamic and uncertainty evaluation of cryogenic turbines, *Twelfth International Symposium on Transport Phenomena (ISTP-12)*, Istanbul, 16–20 July 2000

22. W. Pulkrabek, *Engineering Fundamentals of the Internal Combustion Engine*, 2nd edn. (Prentice Hall, New York, 2004)
23. R.E. Sonntag, C. Borgnakke, G.J. Van Wylen, *Fundamentals of Thermodynamics*, 6th edn. (Wiley, New York, 2002)
24. M. Kanoglu, I. Dincer, Performance assessment of cogeneration plants. Energ. Convers. Manage **50**, 76–81 (2009)
25. M. Kanoglu, S.K. Isik, A. Abusoglu, Performance characteristics of a diesel engine power plant. Energ. Convers. Manage. **46**, 1692–1702 (2005)
26. T.A. Brzustowski, A. Brena, Second law analyses of energy processes. IV – the exergy of hydrocarbon fuels. Trans. Can. Soc. Mech. Eng. **10**(3), 121–128 (1986)
27. E. Barbier, Nature and technology of geothermal energy: a review. Renew. Sust. Energ. Rev. Int. J. **1**(2), 1–69 (1997)
28. R. DiPippo, Second law analysis of flash-binary and multilevel binary geothermal power plants. Geotherm. Res. Council Trans. **18**, 505–510 (1994)
29. M. Kanoglu, Exergy analysis of a dual-level binary geothermal power plant. Geothermics **31**, 709–724 (2004)
30. J. Kestin, Available work in geothermal energy, in *Sourcebook on the Production of Electricity from Geothermal Energy* (U.S. Dept. of Energy, Washington, DC, 1980)
31. M. Kanoglu, Y.A. Cengel, Improving the performance of an existing binary geothermal power plant: a case study. Trans. ASME J. Energ. Res. Technol. **121**(3), 196–202 (1999)
32. A.S. Joshi, Evaluation of cloudiness/haziness factor and its application for photovoltaic thermal (PV/T) system for Indian climatic conditions. Doctoral thesis, IIT Delhi/New Delhi/India, 2006
33. A.S. Joshi, I. Dincer, B.V. Reddy, Energetic and exergetic analyses of a photovoltaic system, in *Proceedings of the Canadian Society for Mechanical Engineering Forum 2008 Conference*. Paper Number 1569103179, Ottawa, 5–8 June 2008
34. A.S. Joshi, I. Dincer, B.V. Reddy, Performance analysis of photovoltaic systems: a review. Renew. Sust. Energ. Rev. **13**(8), 1884–1897 (2009)
35. F.A. Holland, F.A. Watson, S. Devotta, *Thermodynamic Design Data for Heat Pump Systems* (Pergamon Press, Oxford, 1982)
36. ASHRAE American Society of Heating, Refrigerating and Air-Conditioning Engineers, *Handbook of Refrigeration* (ASHRAE, Atlanta, 2006)
37. I. Dincer, M. Kanoglu, *Refrigeration Systems and Applications*, 2nd edn. (Wiley, New York, 2010)
38. R. Barron, *Cryogenic Systems* (Oxford University Press, New York, 1997)
39. K.D. Timmerhaus, T.M. Flynn, *Cryogenic Process Engineering*. The International Cryogenic Monographs Series (Plenum Press, New York, 1989)
40. G. Walker, *Cryocoolers* (Plenum Press, New York, 1983)
41. M. Kanoglu, I. Dincer, M.A. Rosen, Performance analysis of gas liquefaction cycles. Int. J. Energ. Res. **32**(1), 35–43 (2008)
42. M. Kanoglu, Exergy analysis of multistage cascade refrigeration cycle used for natural gas liquefaction. Int. J. Energ. Res. **26**, 763–774 (2002)
43. W.J. Wepfer, R.A. Gaggioli, E.F. Obert, Proper evaluation of available energy for HVAC. ASHRAE Trans. 85, Part I, pp. 214–230 (1979)
44. M. Kanoglu, I. Dincer, M.A. Rosen, Exergy analysis of psychrometric processes for HVAC&R applications, ASHRAE Transactions LB-07-020, *ASHRAE 2007 Annual Meeting*, Long Beach, California, 23–27 June 2007

Index